Symmetry in Quantum Optics Models

Symmetry in Quantum Optics Models

Special Issue Editor

Lucas Lamata

MDPI • Basel • Beijing • Wuhan • Barcelona • Belgrade

Special Issue Editor
Lucas Lamata
Universidad de Sevilla
Spain

Editorial Office
MDPI
St. Alban-Anlage 66
4052 Basel, Switzerland

This is a reprint of articles from the Special Issue published online in the open access journal *Symmetry* (ISSN 2073-8994) in 2019 (available at: https://www.mdpi.com/journal/symmetry/special_issues/Symmetry_Quantum_Optics_Models).

For citation purposes, cite each article independently as indicated on the article page online and as indicated below:

LastName, A.A.; LastName, B.B.; LastName, C.C. Article Title. *Journal Name* **Year**, *Article Number*, Page Range.

ISBN 978-3-03921-858-5 (Pbk)
ISBN 978-3-03921-859-2 (PDF)

Contents

About the Special Issue Editor vii

Lucas Lamata
Symmetry in Quantum Optics Models
Reprinted from: *Symmetry* **2019**, *11*, 1310, doi:10.3390/sym11101310 1

Daniel Braak
Symmetries in the Quantum Rabi Model
Reprinted from: *Symmetry* **2019**, *11*, 1259, doi:10.3390/sym11101259 3

Andreas Lubatsch and Regine Frank
Behavior of Floquet Topological Quantum States in Optically Driven Semiconductors
Reprinted from: *Symmetry* **2019**, *11*, 1246, doi:10.3390/sym11101246 16

Ricardo Puebla, Giorgio Zicari, Iñigo Arrazola, Enrique Solano, Mauro Paternostro and Jorge Casanova
Spin-Boson Model as A Simulator of Non-Markovian Multiphoton Jaynes-Cummings Models
Reprinted from: *Symmetry* **2019**, *11*, 695, doi:10.3390/sym11050695 32

Francisco A. Cárdenas-López, Guillermo Romero, Lucas Lamata, Enrique Solano and Juan Carlos Retamal
Parity-Assisted Generation of Nonclassical States of Light in Circuit Quantum Electrodynamics
Reprinted from: *Symmetry* **2019**, *11*, 372, doi:10.3390/sym11030372 53

Jorge A. Anaya-Contreras, Arturo Zúñiga-Segundo and Héctor M. Moya-Cessa
Quasiprobability Distribution Functions from Fractional Fourier Transforms
Reprinted from: *Symmetry* **2019**, *11*, 344, doi:10.3390/sym11030344 73

About the Special Issue Editor

Lucas Lamata (Prof.) is an Associate Professor of Theoretical Physics at Universidad de Sevilla, Spain, affiliated to the Department of Atomic, Molecular, and Nuclear Physics of the Faculty of Physics.

His research up to now has focused on quantum optics and quantum information, including pioneering proposals for quantum simulations of relativistic quantum mechanics, fermionic systems, and spin models, with trapped ions and superconducting circuits. He is also interested in new approaches to quantum simulation, as with his novel concept of embedding quantum simulators, and in the emulation of biological behaviors with quantum controllable systems, in the research line that he calls quantum biomimetics, developed at his previous position at the University of the Basque Country. He is also analyzing the possibility of combining artificial intelligence and machine learning protocols with quantum devices. He enjoys working with experimentalists and has made proposals and participated in 15 experiments in collaboration with up to 15 prominent experimental groups in quantum science, with trapped ions, electrons in Penning traps, superconducting circuits, quantum photonics, and nuclear magnetic resonance. Up to 16 of his theoretical proposals for implementations have been carried out in experiments by top-flight groups.

Before working in Sevilla, he was at Bilbao, in the QUTIS Group led by Prof. Enrique Solano, first as a Marie Curie postdoctoral fellow and subsequently in a Ramón y Cajal position and a Staff Scientist position. Before this, he was a Humboldt Fellow and a Max Planck postdoctoral fellow for three and a half years at the Max Planck Institute for Quantum Optics in Garching, Germany, working in Prof. Ignacio Cirac Group. Previously, he carried out his PhD at CSIC, Madrid, and Universidad Autónoma de Madrid (UAM), supervised by Prof. Juan León. His PhD thesis was awarded with the First Extraordinary Prize for a PhD in Physics in 2007 in UAM, out of more than 30 theses. He has more than 16 years of research experience in centers in Spain and Germany, having performed research as well with scientific collaborations in several one- or two-week stays in centers of all continents as Harvard University, ETH Zurich, University of California Berkeley, Google Santa Barbara, University of California Santa Barbara, Google LA, Shanghai University, Tsinghua University, Macquarie University, University of Bristol, Walther-Meissner Institut Garching, University of KwaZulu-Natal, IQOQI Innsbruck, and Universidad de Santiago de Chile, among others.

He has published and submitted about 100 articles in international refereed journals, including: 1 in *Nature*, 1 in *Reviews of Modern Physics*, 1 in *Advances in Physics: X*, 3 in *Nature Communications*, 2 in *Physical Review X*, 1 in *APL Photonics*, 2 in *Advanced Quantum Technologies*, 1 in *Quantum Science and Technology*, and 19 in *Physical Review Letters*, two of them as Editor's Suggestion. Overall, he has published 24 articles in Nature Publishing Group journals, 45 in American Physical Society journals, and 26 articles in first decile (D1) journals.

His H-index according to Google Scholar is 31, with more than 3500 citations. His i10 index is of 65.

He is or has been PI/Co-PI of several European, USA, and Spanish national grants, serves in the editorial board of six prestigious scientific journals, and has been guest editor of three Special Issues for three different journals. He is also a reviewer for about 50 indexed scientific journals, including 10 journals with an Impact Factor larger than 8, according to 2018 SCI.

symmetry

Editorial

Symmetry in Quantum Optics Models

Lucas Lamata

Departamento de Física Atómica, Molecular y Nuclear, Universidad de Sevilla, Apartado 1065, 41080 Sevilla, Spain; lucas.lamata@gmail.com

Received: 16 October 2019; Accepted: 17 October 2019; Published: 18 October 2019

This editorial introduces the successful invited submissions [1–5] to a Special Issue of Symmetry on the subject area of "Symmetry in Quantum Optics Models".

Quantum optics techniques can be regarded as the physical background of quantum technologies. These techniques are most often enhanced by symmetry considerations, which can simplify calculations as well as offer new insight into the models.

This Special Issue includes the novel techniques and tools for Quantum Optics Models and Symmetry, such as:

- Quasiprobability distribution functions employing fractional Fourier transforms [1].
- Ultrastrong coupling regime combined with parity symmetry for the nonclassical state of light generation [2].
- Employment of quantum optics models as the spin-boson system for simulating another quantum optics platform, as non-Markovian multi-photon Jaynes–Cummings models [3].
- Floquet topological techniques for analyzing optically driven semiconductors [4].
- Symmetries of the quantum Rabi model for its analysis in all possible parameter regimes [5].

Response to our call had the following statistics:

- Submissions (5);
- Publications (5);
- Article types: Research Article (5).

Authors' geographical distribution (published papers) is:

- China (2)
- Spain (2)
- Germany (2)
- Mexico (1)
- UK (1)
- Chile (1)
- USA (1)

Published submissions are related to the aforementioned techniques and tools, and represent a selection of current topics in quantum optics models and their symmetries.

We found the edition and selections of papers for this book very inspiring and rewarding. We also thank the editorial staff and reviewers for their efforts and help during the process.

Conflicts of Interest: The author declares no conflict of interest.

References

1. Anaya-Contreras, J.; Zúñiga-Segundo, A.; Moya-Cessa, H. Quasiprobability Distribution Functions from Fractional Fourier Transforms. *Symmetry* **2019**, *11*, 344. [CrossRef]
2. Cárdenas-López, F.; Romero, G.; Lamata, L.; Solano, E.; Retamal, J. Parity-Assisted Generation of Nonclassical States of Light in Circuit Quantum Electrodynamics. *Symmetry* **2019**, *11*, 372. [CrossRef]
3. Puebla, R.; Zicari, G.; Arrazola, I.; Solano, E.; Paternostro, M.; Casanova, J. Spin-Boson Model as A Simulator of Non-Markovian Multiphoton Jaynes-Cummings Models. *Symmetry* **2019**, *11*, 695. [CrossRef]
4. Lubatsch, A.; Frank, R. Behavior of Floquet Topological Quantum States in Optically Driven Semiconductors. *Symmetry* **2019**, *11*, 1246. [CrossRef]
5. Braak, D. Symmetries in the Quantum Rabi Model. *Symmetry* **2019**, *11*, 1259. [CrossRef]

symmetry MDPI

Article

Symmetries in the Quantum Rabi Model

Daniel Braak

Max-Planck Institute for Solid State Research, Heisenbergstr. 1, 70569 Stuttgart, Germany; d.braak@fkf.mpg.de

Received: 21 September 2019; Accepted: 6 October 2019; Published: 9 October 2019

Abstract: The quantum Rabi model is the simplest and most important theoretical description of light–matter interaction for all experimentally accessible coupling regimes. It can be solved exactly and is even integrable due to a discrete symmetry, the $\mathbb{Z}_2$ or parity symmetry. All qualitative properties of its spectrum, especially the differences to the Jaynes–Cummings model, which possesses a larger, continuous symmetry, can be understood in terms of the so-called "G-functions" whose zeroes yield the exact eigenvalues of the Rabi Hamiltonian. The special type of integrability appearing in systems with discrete degrees of freedom is responsible for the absence of Poissonian level statistics in the spectrum while its well-known "Juddian" solutions are a natural consequence of the structure of the G-functions. The poles of these functions are known in closed form, which allows drawing conclusions about the global spectrum.

Keywords: light–matter interaction; integrable systems; global spectrum

---

1. Introduction

The spectacular success of quantum optics [1] is based to a considerable extent on the fact that the light quanta do not interact among themselves. On the other hand, the interaction of quantized radiation with matter is quite complicated because even the simplest model, an atomic two-level system coupled to a single radiation mode via a dipole term, does not conserve the excitation number. This model, the quantum Rabi model (QRM) [2–4], is of central importance as basically all experimental observations in the field can be traced to a variant of it [5]. The QRM Hamiltonian reads

$$H_R = \omega a^\dagger a + g\sigma_x(a + a^\dagger) + \Delta\sigma_z. \tag{1}$$

Here, $a^\dagger$ and a are the creation and annihilation operators of the bosonic mode and energy is measured in units of frequency ($\hbar = 1$). 2Δ denotes the energy splitting of the two-level system, which is coupled linearly to the electric field ($\sim (a + a^\dagger)$) with interaction strength g. The QRM has just two degrees of freedom, one continuous (the radiation mode) and one discrete (the two-level system), described by Pauli matrices σ_z, σ_x. Even better known than the QRM is a famous approximation to it, the Jaynes–Cummings model (JCM),

$$H_{JC} = \omega a^\dagger a + g(\sigma^+ a + \sigma^- a^\dagger) + \Delta\sigma_z, \tag{2}$$

with $\sigma^\pm = (\sigma_x \pm i\sigma_y)/2$. In this model, the "counter-rotating terms" $g(\sigma^+ a^\dagger + \sigma^- a)$ are missing, so that it conserves the excitation number $\hat{C} = a^\dagger a + \sigma^+\sigma^-$ and can be solved analytically in closed form [4]. The QRM, including these terms, was long considered to be unsolvable by analytical means and also non-integrable [6], until its exact solution was discovered [7].

The JCM provides very good agreement with experiments in atom optics where the dipole coupling strength is many orders of magnitude smaller than the mode frequency. Its characteristic feature manifests itself for example in the vacuum Rabi splitting, observable if the coupling is larger than the cavity decay rates. This was achieved in an experiment from 1992 with a ratio $g/\omega = 10^{-8}$ between dipole coupling and mode frequency [8]. Since then, there has been tremendous progress in the experimental techniques to enhance the coupling strength between light and matter within a wide range of different platforms, ranging from cavity quantum electrodynamics, using optical and microwave frequencies, to circuit QED, which implements the radiation mode in a transmission line, while the coupled two-level system is realized in various ways, e.g. via superconducting qubits or quantum dots, as excitonic or intersubband polaritons [9,10]. Within the last 27 years, the ratio g/ω has been raised by eight orders of magnitude, finally reaching the so-called deep strong coupling regime (DSC) [11], $g \sim \omega$ within a circuit QED framework [12].

For these coupling strengths, the JCM is no longer applicable and gives even qualitatively wrong results. Already for $0.1 \lesssim g/\omega \lesssim 0.3$, called the perturbative ultra-strong coupling regime (pUSC) [10], there are measurable deviations [13], although these can still be accounted for by the Bloch–Siegert Hamiltonian [14,15], a solvable extension of the JCM. For $g/\omega > 0.3$, one enters the non-perturbative ultra-strong coupling regime (USC), where also the Bloch–Siegert Hamiltonian fails.

Part of the interest in the USC and DSC regimes originates in the natural identification of the two-level system with a qubit, the building block of quantum information theory [16]. The strong coupling between the qubit and light field allows for novel technologies such as nondestructive readout [17] and remote entanglement [18] besides the possibility to implement quantum error correcting codes [19]. However, the strong coupling regimes are also fascinating from the viewpoint of fundamental research, because the light–matter system behaves in unexpected and sometimes counter-intuitive ways: the vacuum state contains virtual photons [20] and in the DSC the Purcell effect disappears [21] while the standard collapse and revival dynamics of the two-level system becomes dominated by the mode frequency [11].

2. The Rotating-Wave Approximation and Its Symmetry

These developments have renewed the interest in the analytical understanding of the QRM beyond a brute-force diagonalization of the Hamiltonian in a truncated, finite-dimensional Hilbert space. To this end, several improvements of the rotating-wave approximation underlying the JCM have been proposed [22–24] which should be reliable even for strong coupling. However, all methods, while being quantitatively in reasonable agreement with the numerical diagonalization, deviate qualitatively from it by predicting degeneracies absent in the true spectrum of the QRM.

The JCM reproduces the exact spectrum with great accuracy almost up to the first level crossings (counted from the left of the spectral graph in Figure 1), which is a true crossing, actually the first Juddian solution [25]. However, the next crossings of the JCM which appear for $g \gtrsim 0.5$ (marked with small green circles in Figure 1) are avoided in the QRM. The reason is the much larger symmetry of the JCM compared to the QRM. Because $[H_{JC}, \hat{C}] = 0$, each eigenstate of the JCM is also an eigenstate of $\hat{C}$ and labeled by corresponding eigenvalue $\lambda_{\hat{C}} = 0, 1, 2, \ldots$ of $\hat{C}$. The eigenspace of $\hat{C}$ with fixed $\lambda_{\hat{C}} = n$ for $n \geq 1$ is two-dimensional while the ground state of the JCM (for sufficiently small g) is the unique state $|\mathrm{vac}\rangle = |0\rangle \otimes |\downarrow\rangle$ with $\lambda_{\hat{C}} = 0$. In other words, the Hilbert space $\mathcal{H} = L^2[\mathbb{R}] \otimes \mathbb{C}^2$ decays into a direct sum of dynamically invariant subspaces

$$\mathcal{H} = |\mathrm{vac}\rangle \oplus \sum_{n=1}^{\infty} \mathcal{H}_n, \tag{3}$$

where each $\mathcal{H}_n$ is two-dimensional.

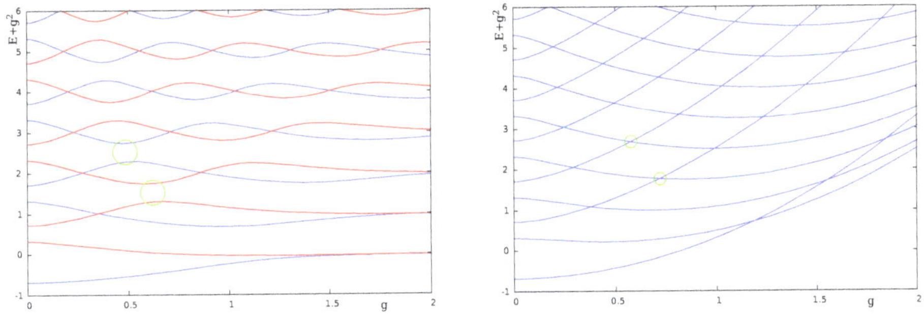

Figure 1. (**Left**) The QRM spectrum for $\omega = 1, \Delta = 0.7$ as function of the coupling constant g. Instead of the energy, the spectral parameter $x = E + g^2$ is displayed on the ordinate. States with negative (positive) parity are displayed in blue (red). Within the same parity subspace all level crossings are avoided (green circles). (**Right**) The Jaynes–Cummings-spectrum for the same parameters. In this case, corresponding states do cross due to the enhanced symmetry of the JCM (small green circles).

Thus, the eigenstates with $\lambda_{\hat{C}} > 0$ can be labeled by two quantum numbers, the first gives the eigenvalue of $\hat{C}$, and the second takes just two values 0 and 1, corresponding to the two states in $\mathcal{H}_n$, forming the so-called Jaynes–Cummings doublets. As the infinitely many subspaces are dynamically disconnected for all values of g, the energies $E_{n,j}$ and $E_{m,j'}$ may become degenerate whenever $n \neq m$. The two crossings selected in Figure 1 are degeneracies between the JC-states $|1,1\rangle$ and $|3,0\rangle$ and between $|2,1\rangle$ and $|4,0\rangle$, respectively. In contrast to these degeneracies that are lifted by the counter-rotating terms, the crossings between the JC-states $|1,1\rangle$ and $|2,0\rangle$ are also present in the spectral graph of the QRM.

Because the algebra $\mathcal{A} = \langle \mathbb{1}, \hat{C}, \hat{C}^2, \dots \rangle$ generated by $\hat{C}$ is infinite dimensional, the operators

$$\hat{U}(\phi) = \exp(i\phi\hat{C}) = \sum_{n=0}^{\infty} \frac{(i\phi\hat{C})^n}{n!} = e^{i\phi a^\dagger a} \otimes \begin{pmatrix} e^{i\phi} & 0 \\ 0 & 1 \end{pmatrix} \tag{4}$$

are linearly independent for all $0 \leq \phi < 2\pi$. However, because the spectrum of $a^\dagger a$ is integer-valued, we have $\hat{U}(2\pi) = \mathbb{1}$ and the $\hat{U}(\phi)$ form an infinite dimensional representation of the continuous compact group $U(1)$ in $\mathcal{H}$ with composition law $\hat{U}(\phi_1)\hat{U}(\phi_2) = \hat{U}(\phi_1 + \phi_2)$. We have for any ϕ the relation $U^\dagger(\phi)H_{JC}U(\phi) = H_{JC}$, as

$$U^\dagger(\phi)aU(\phi) = e^{i\phi}a, \quad U^\dagger(\phi)a^\dagger U(\phi) = e^{-i\phi}a^\dagger, \quad U^\dagger(\phi)\sigma^\pm U(\phi) = e^{\mp i\phi}\sigma^\pm, \quad 0 \leq \phi < 2\pi. \tag{5}$$

This means that the "rotating" interaction term $a^\dagger\sigma^- + a\sigma^+$ is invariant for the whole group but the "counter-rotating" term $a^\dagger\sigma^+ + a\sigma^-$ is invariant only for $\phi = \pi$. Indeed, the set $\{\mathbb{1}, \hat{U}(\pi)\}$ forms a discrete subgroup of $U(1)$. Because

$$\hat{U}(\pi) = -(-1)^{ia^\dagger a} \otimes \sigma_z = -\hat{P}, \quad \text{with} \quad \hat{P}^2 = \mathbb{1}, \tag{6}$$

it is the group with two elements $\{\mathbb{1}, \hat{P}\} = \mathbb{Z}/2\mathbb{Z} \equiv \mathbb{Z}_2$ (the sign of the "parity" operator $\hat{P}$ is chosen here to conform with the convention in [7]). The QRM is invariant under the finite group $\mathbb{Z}_2$, $\hat{P}H_R\hat{P} = H_R$.

The character group of $U(1)$ is $\mathbb{Z}$, therefore each one-dimensional irreducible representation of $U(1)$ is labeled by an integer $n \in \mathbb{Z}$. In the representation in Equation (4), the space $\mathcal{H}_n$ spanned by the vectors $|n-1\rangle \otimes |\uparrow\rangle$ and $|n\rangle \otimes |\downarrow\rangle$ for $n \geq 1$ is invariant and $\hat{U}(\phi)$ acts on it as $e^{in\phi}\mathbb{1}_2$. Therefore, the decomposition in Equation (3) corresponds to the irreducible representations of $U(1)$ in $\mathcal{H}$ for integers $n \geq 0$ and the spectral problem for the JCM reduces to the diagonalization of 2×2-matrices in the spaces $\mathcal{H}_n$ [4]. If one parameter of the model is varied, say the coupling g, the spaces $\mathcal{H}_n$ do not change, only the eigenvectors $|n, j\rangle \in \mathcal{H}_n$, $j = 0, 1$ and the eigenenergies $E_{n,j}$. The spectral graph as function of g consists of infinitely many ladders with two rungs, intersecting in the E/g-plane as shown for $2\Delta = \omega$ in Figure 2.

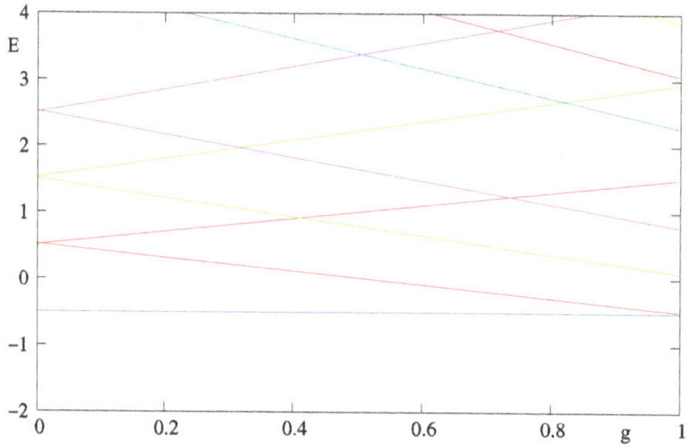

Figure 2. The JCM spectrum at resonance $2\Delta = \omega = 1$ as a function of g. Each color corresponds to an invariant subspace $\mathcal{H}_n$. The state $|\text{vac}\rangle = |0\rangle \otimes |\downarrow\rangle$ spans the (trivial) irreducible representation of $U(1)$ with character 0.

We find that the *continuous* symmetry of the JCM allows to classify the eigenstates according to infinitely many irreducible representations, thereby effectively eliminating the *continuous* (bosonic) degree of freedom, the radiation mode. The remaining discrete degree of freedom (the two-level system) has a two-dimensional Hilbert space and, after application of the $U(1)$-symmetry, the Hamiltonian acts non-trivially only in the two-dimensional $\mathcal{H}_n$. The JCM possesses an additional conserved quantity, $\hat{C}$, besides the Hamiltonian H_{JC}. As it has two degrees of freedom, it is therefore integrable according to the classical criterion by Liouville [26], because the number of phase-space functions (operators) in involution equals the number of degrees of freedom.

What about the QRM? We have $[\hat{P}, H_R] = 0$, but the associated symmetry is discrete and has only two irreducible representations, corresponding to the eigenvalues $\lambda_{\hat{P}} = \pm 1$ of $\hat{P}$. It follows that the Hilbert space decomposes into the direct sum

$$\mathcal{H} = \mathcal{H}_+ \oplus \mathcal{H}_-. \tag{7}$$

Both $\mathcal{H}_\pm$ are infinite dimensional and the spectral problem appears as complicated as before. However, in each parity subspace (usually called parity chain [11]), the discrete degree of freedom has been eliminated and only the continuous degree of freedom remains. According to the standard reasoning, a conservative system with only one degree of freedom is integrable. From this point of view, advocated in [7], the QRM is integrable because the *discrete* $\mathbb{Z}_2$-symmetry has eliminated the *discrete* degree of freedom. This is only possible because the number of irreducible representations of $\mathbb{Z}_2$ matches precisely the dimension of the Hilbert space $\mathbb{C}^2$ of the two-level system. Other models with one continuous and one discrete degree of freedom such as the Dicke models with Hilbert space $L^2[\mathbb{R}] \otimes \mathbb{C}^n$ are not integrable according to this criterion, because their $\mathbb{Z}_2$-symmetry is not sufficient to reduce the model to a single continuous degree of freedom if $n > 2$ [27]. On the other hand, the continuous symmetry introduced by the rotating-wave approximation is so strong that it renders the Dicke model integrable for all n [28].

The criterion on quantum integrability proposed in [7] is especially suited to systems with a single continuous and several discrete degrees of freedom and states then that a system is quantum integrable if each eigenstate can be labeled uniquely by a set of quantum numbers $|\psi\rangle = |n; m_1, m_2, \ldots\rangle$ where $0 \leq n < \infty$ corresponds to the continuous degree of freedom and the number of different tuples $\{m_1, m_2, \ldots\}$ equals the dimension d of the Hilbert space belonging to the discrete degrees of freedom. This unique labeling allows then for degeneracies between states belonging to different tuples $\{m_1, m_2, \ldots\}$, which characterize the different decoupled subspaces $\mathcal{H}_{\{m_1,m_2,\ldots\}}$. Within the space $\mathcal{H}_{\{m_1,m_2,\ldots\}}$, which is infinite dimensional and isomorphic to $L^2[\mathbb{R}]$, the states are labeled with the single number n and level crossings are usually avoided between states $|n; m_1, m_2, \ldots\rangle$ and $|n'; m_1, m_2, \ldots\rangle$ if no continuous symmetry is present. This happens in the QRM, where the spectral graph is composed of two ladders each with infinitely many rungs (see Figure 1). The situation is in some sense dual to the JCM, where we have infinitely many intersecting ladders with two rungs. The stronger symmetry of the JCM renders it therefore superintegrable [29].

With a stronger symmetry, more degeneracies are to be expected. Especially going from a discrete to a continuous symmetry by applying the rotating-wave approximation inevitably introduces unphysical level crossings in the spectral graph. This applies especially to those methods which apply the rotating-wave approximation on top of unitary transformations such as the GRWA [22–24]. In Figure 3, it is seen that the spectral graph provided by the GRWA indeed reproduces correctly all level crossings of the QRM in the E/g-plane but exhibits unphysical level crossings in the E/Δ-plane.

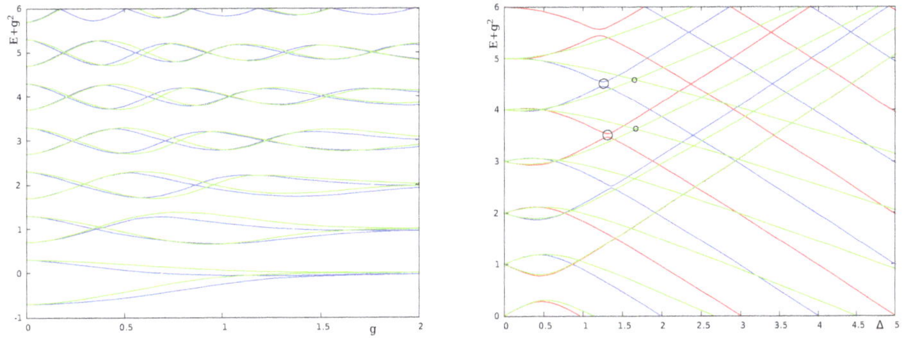

Figure 3. (**Left**) The QRM spectrum (blue) and the approximation by the GRWA (green) as function of g for $\Delta = 0.7$. The GRWA reproduces the qualitative properties of the spectral graph also for large coupling. (**Right**) The QRM and GRWA spectra as function of Δ for $g = 0.25$. The blue (red) level lines correspond to negative (positive) parity in the QRM. In this case, the GRWA shows level crossings (small black circles) where the QRM has none (black circles) because there are no degeneracies for fixed parity. All apparent degeneracies of the QRM within the same parity chain are narrow avoided crossings.

3. Integrability of Systems with Less Than Two Continuous Degrees of Freedom

The notion of integrability in quantum systems is still controversial [30] and based mainly either on the Bethe ansatz [31] or on the statistical criterion by Berry and Tabor [32]. While it was demonstrated by Amico *et al.* [6] and Batchelor and Zhou [33] that the QRM is not amenable to the Bethe ansatz, its level statistics deviate markedly from the Poissonian form for the average distance $\Delta E = E_{n+1} - E_n$ between energy levels. According to Berry and Tabor [32], the distribution of ΔE in a quantum integrable system should read $P(\Delta E) \sim \exp(-\Delta E/\langle \Delta E \rangle)$, where $\langle \Delta E \rangle$ is the average level distance in a given energy window. This distribution is not present in the QRM [34], whose level distances are shown in Figure 4 up to $n \sim 5000$.

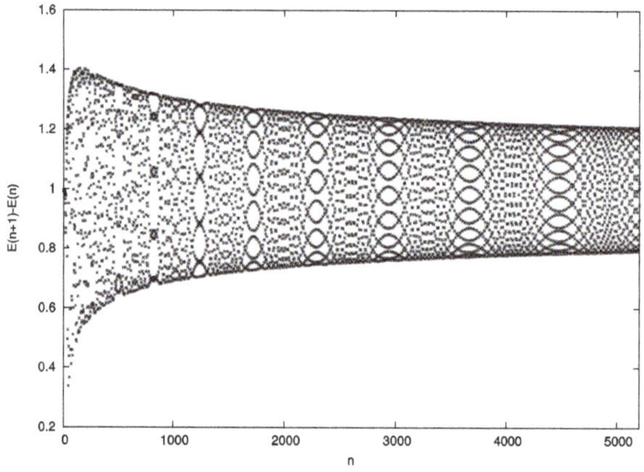

Figure 4. The distribution of level distances $\Delta E = E_{n+1} - E_n$ of the QRM for positive parity as function of the level number n. Parameters are $\omega = 1, g = \Delta = 5$. A clear deviation from the exponential law predicted in [32] is visible.

Due to this deviation from the expected behavior for integrable systems and likewise from the Wigner surmise [35], it was unclear whether the QRM belongs to the integrable or chaotic systems [34]. If the QRM is integrable as argued above, why does the Berry–Tabor criterion not apply? The reason lies in the fact that this criterion has been derived for classically integrable systems with N continuous degrees of freedom, which can be quantized with the Bohr–Sommerfeld method. In this case the energy eigenvalues are labeled by N integers n_j. The classical Hamiltonian can be written as an in general non-linear function of N action variables $I_1, \ldots, I_N$, $H = f(I_1, \ldots, I_N)$. Then the quantized energies read

$$E_{n_1,\ldots,n_N} = f(\hbar(n_1 + \alpha_1/4), \ldots, \hbar(n_N + \alpha_N/4)) = \tilde{f}(n_1, \ldots, n_N), \tag{8}$$

where the α_j are Maslov indices. The level distance distribution follows then from the statistics of vectors $(n_1, \ldots, n_N)$ with integer entries belonging to the energy shell $E \leq \tilde{f}(n_1, \ldots, n_N) \leq E + \delta E$. This is shown for $N = 2$ in Figure 5.

Berry and Tabor showed that the occurrences of the $(n_1, \ldots, n_N)$ in the shell $[E, E + \delta E]$ are essentially uncorrelated provided $\tilde{f}(n_1; \ldots, n_N)$ is a non-linear function of its arguments and $N \geq 2$. $\tilde{f}$ is linear for linearly coupled harmonic oscillators [32] and in this case the level statistics is not Poissonian. The criterion applies thus only to systems with at least two continuous degrees of freedom. If one of the degrees of freedom is discrete, the corresponding action variable takes only finitely many values. This has the same effect as a linear $\tilde{f}$. A deviation from Poissonian statistics would therefore be expected even if the QRM would be the quantum limit of a classically integrable system. However, this is not the case. The weak symmetry of the QRM may have a counterpart in the classical limit but then it would not suffice to make the classical model (which must have at least two continuous degrees of freedom) integrable. The QRM is integrable only as a genuine quantum model. The Hilbert space of the quantum degree of freedom must not be larger than two—otherwise the model becomes non-integrable similar to the Dicke model [27].

Figure 6 shows on the left the spectral graph of the Dicke model for three qubits (which is also exactly solvable by the method described in the next section) with Hamiltonian

$$H_D = a^\dagger a + 2g(a + a^\dagger)\hat{J}_z + 2\Delta\hat{J}_x, \tag{9}$$

where $\hat{J}_z$ an $\hat{J}_x$ are generators of $SU(2)$ in the spin-$\frac{3}{2}$ representation. The QRM spectrum is depicted on the right. It is apparent that most of the regular features of the Rabi spectrum are absent in the Dicke spectrum, although it has the same $\mathbb{Z}_2$-symmetry.

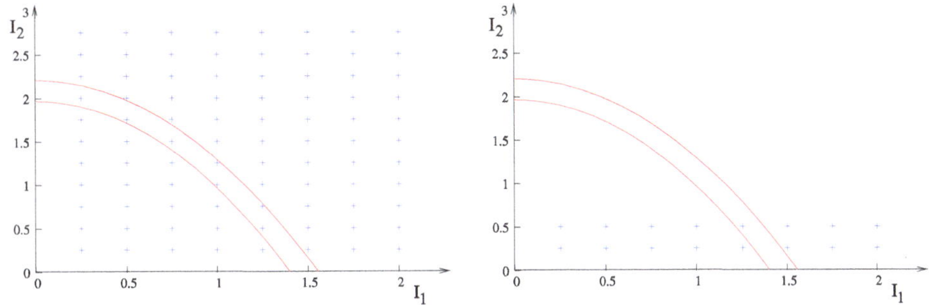

Figure 5. (**Left**) The energy shell $[E, E + \delta E]$ (red lines) contains the integer-valued vectors (n_1, n_2) (blue crosses) belonging to the quantization of the action variables $I_1 = \hbar(n_1 + \alpha_1/4)$ and $I_2 = \hbar(n_2 + \alpha_2/4)$. The distance of adjacent energies $\tilde{f}(n_1, n_2) - \tilde{f}(n_1', n_2')$ is statistically unrelated for large quantum numbers if $\tilde{f}$ is non-linear. (**Right**) If the second action variable I_2 can take only two values as would be the case for a discrete degree of freedom with $\dim\mathcal{H} = 2$, the average level distance is the same as for linear $\tilde{f}$ and Poisson statistics does not apply.

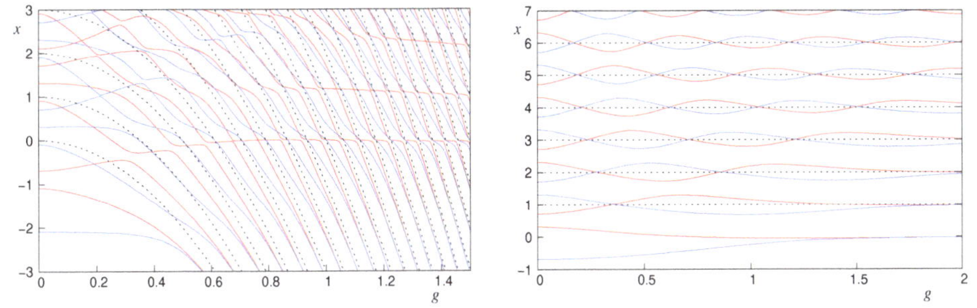

Figure 6. (**Left**) The Dicke spectrum for spin $3/2$ and $\Delta = 0.7$ as function of g. The spectral parameter is $x = E + g^2/3$, making the "baselines of the first kind" [27] horizontal while the baselines of the second kind are given as dashed lines. (**Right**) The QRM spectrum at $\Delta = 0.7$ for comparison ($x = E + g^2$). All level crossings are located on the horizontal baselines with $x = \text{const}$.

4. The Global Spectrum of the QRM

As shown in Figure 6, the spectral graph of the QRM has an intriguingly simple structure. The level lines cross only for different parity on the so-called "baselines" with $x = n$ for $n = 1, 2, \ldots$. Moreover, the degenerate states are quasi-exact solutions whose wave function can be expressed through polynomials [25]. These features can be explained in a unified way by the properties of the spectral determinant or G-function of the QRM, $G_\pm(x)$, whose zeroes give the exact eigenvalues of the Hamiltonian in each parity chain [7]. The G-function of the QRM is given as the following function ($\omega = 1$),

$$G_\pm(x;g,\Delta) = \sum_{n=0}^{\infty} K_n(x) \left[1 \mp \frac{\Delta}{x-n}\right] g^n, \tag{10}$$

where the $K_n(x)$ are defined recursively,

$$nK_n = f_{n-1}(x)K_{n-1} - K_{n-2}, \tag{11}$$

with

$$f_n(x) = 2g + \frac{1}{2g}\left(n - x + \frac{\Delta^2}{x-n}\right), \tag{12}$$

and initial condition $K_0 = 1, K_1(x) = f_0(x)$. Note that $G_-(x;g,\Delta) = G_+(x;g,-\Delta)$. The G-functions can be written in terms of confluent Heun functions [36], namely

$$G_\pm(x) = \left(1 \mp \frac{\Delta}{x}\right) H_c(\alpha,\gamma,\delta,p,\sigma;1/2) - \frac{1}{2x} H_c'(\alpha,\gamma,\delta,p,\sigma;1/2). \tag{13}$$

where $H_c'(\alpha,\gamma,\delta,p,\sigma;z)$ denotes the derivative of $H_c(\alpha,\gamma,\delta,p,\sigma;z)$ with respect to z. The parameters are given as [37],

$$\alpha = -x, \quad \gamma = 1 - x, \quad \delta = -x,$$
$$p = -g^2, \quad \sigma = x(4g^2 - x) + \Delta^2.$$

From Equations (11) and (12) one may deduce that $G_\pm(x)$ has simple poles at $x = 0, 1, 2, \dots$ and therefore its zeroes are usually not located at integers but pinched between the poles. $G_\pm(x;g,\Delta)$ can be written as

$$G_\pm(x;g,\Delta) = \tilde{G}_\pm(x;g,\Delta) + \sum_{n=0}^{\infty} \frac{h_n^\pm(\Delta,g)}{x-n}, \tag{14}$$

where $\tilde{G}_\pm(x;g,\Delta)$ is analytic in x and $\tilde{G}_\pm(x;g,\Delta) \approx e^{2g^2}2^{-x}$ for small Δ. The coefficients $h_n^\pm(\Delta,g)$ vanish for $\Delta = 0$. Indeed, the sign of $h_n^\pm$ determines whether the zero of $G_\pm(x)$ in the vicinity of $x = n$ is located to the right or to the left of n in the two adjacent intervals $n-1 < x < n$ and $n < x < n+1$. This leads to the following conjecture about the distribution of zeroes of $G_\pm(x)$:

Conjecture 1. *The number of zeros in each interval $[n, n+1]$, $n \in \mathbb{N}_0$ is restricted to be 0, 1, or 2. Moreover, an interval $[n, n+1]$ with two roots of $G_\pm(x) = 0$ can only be adjacent to an interval with one or zero roots; in the same way, an empty interval can never be adjacent to another empty interval.*

Figure 7 shows on the left $G_+(x)$ for $g = 0.4$ and $\Delta = 1$ together with the analytic approximation for $\Delta = 0$. The G-conjecture appears to be valid for arbitrary Δ as is shown on the right of Figure 7.

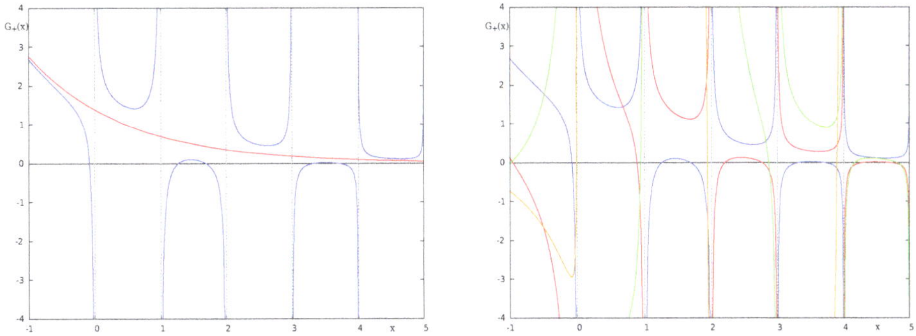

Figure 7. (**Left**) $G_+(x)$ and its entire approximation $G_+(x;0.4,0)$ for $\Delta = 1$. (**Right**) $G_+(x)$ for $\Delta = 1$ (blue), 2 (red), 4 (green) and 7 (orange).

The G-conjecture has not yet been proven in general, although it is possible to prove it for small Δ, which, however, is equivalent with perturbation theory in the operator σ_z, the natural bounded perturbation of the QRM, in contrast to the unbounded coupling operator $\sigma_x(a + a^\dagger)$. Preliminary steps in the direction of a general proof are given in [38].

Assuming the G-conjecture to be valid also for generalizations of the QRM showing the "spectral collapse" phenomenon [39] allows deriving the continuous spectrum at the collapse point [40], where numerical methods fail due to the proliferation of low-lying eigenstates.

The G-functions are derived by using the analyticity properties of the eigenfunctions in the Bargmann space, which also explains the degenerate spectrum (the Juddian solutions) in a natural way simply by doing a Frobenius analysis of the relevant differential equations in the complex domain [7]. Let H_+ denote H_R restricted to the subspace with positive parity. In the Bargmann representation, the Schrödinger equation $(H_+ - E)\psi(z) = 0$ is equivalent to a linear but non-local differential equation in the complex domain,

$$z\frac{\mathrm{d}}{\mathrm{d}z}\psi(z) + g\left(\frac{\mathrm{d}}{\mathrm{d}z} + z\right)\psi(z) = E\psi(z) - \Delta\psi(-z). \tag{15}$$

With the definition $\psi(z) = \phi_1(z)$ and $\psi(-z) = \phi_2(z)$, we obtain the coupled local system,

$$(z+g)\frac{\mathrm{d}}{\mathrm{d}z}\phi_1(z) + (gz - E)\phi_1(z) + \Delta\phi_2(z) = 0, \tag{16}$$

$$(z-g)\frac{\mathrm{d}}{\mathrm{d}z}\phi_2(z) - (gz + E)\phi_2(z) + \Delta\phi_1(z) = 0. \tag{17}$$

This system has two regular singular points at $z = \pm g$ and an (unramified) irregular singular point of s-rank two at $z = \infty$ [36]. With $x = E + g^2$, the Frobenius exponents of $\phi_1(z)$ at the regular singular point g $(-g)$ are $\{0, 1+x\}$ $(\{0, x\})$, while for $\phi_2(z)$ the exponents at g $(-g)$ are $\{0, x\}$ $(\{0, 1+x\})$ [41]. The eigenfunctions have to be analytic in all of $\mathbb{C}$, therefore the spectrum of H_+ separates naturally in a *regular* part with $x \notin \mathbb{N}_0$ and the *exceptional* part with $x \in \mathbb{N}_0$ [7]. For general values of g, Δ, the exceptional part is empty and all eigenstates are regular.

For $x \notin \mathbb{N}_0$, one of the two linearly independent solutions for $\phi_1(z)$ is not admissible. That means that $\phi_1(z)$ will in general develop a branchpoint with exponent $1 + x$ at $z = g$ even if it is analytic with exponent 0 at $z = -g$. $G_+(x)$ vanishes at those x for which both $\phi_1(z)$ and $\phi_2(z)$ have exponent 0 at g and

$-g$, rendering $\psi(z)$ analytic. To find the exceptional spectrum, we define $y = z + g$, $\phi_{1,2} = e^{-gy+g^2}\bar{\phi}_{1,2}$. Then,

$$y\frac{\mathrm{d}}{\mathrm{d}y}\bar{\phi}_1 = x\bar{\phi}_1 - \Delta\bar{\phi}_2, \tag{18}$$

$$(y-2g)\frac{\mathrm{d}}{\mathrm{d}y}\bar{\phi}_2 = (x - 4g^2 + 2gy)\bar{\phi}_2 - \Delta\bar{\phi}_1. \tag{19}$$

A Frobenius solution with exponent 0 at $y = 0$ may be written as $\bar{\phi}_2(y) = \sum_{n=0}^{\infty} K_n(x)y^n$. Then, the integration of Equation (18) yields

$$\bar{\phi}_1(y) = cy^x - \Delta\sum_{n=0}^{\infty} K_n(x)\frac{y^n}{n-x}. \tag{20}$$

If $x \notin \mathbb{N}_0$, c must be zero. This determines $\bar{\phi}_1(z)$ uniquely in terms of $\bar{\phi}_2(z)$ and the K_n are thus given by the recurrence in Equation (11), leading to the regular spectrum.

Now, let us assume $x = n \in \mathbb{N}_0$. In this case, a solution for $\bar{\phi}_2(y)$ analytic at $y = 0$ may be written as $\bar{\phi}_2(y) = \sum_{m=n+1}^{\infty} K_m(x)y^m$ because $x + 1 > 0$ [41]. In this case, the c in (20) need not to be zero, the K_m satisfy still the recurrence in Equation (11), but with initial condition $K_n = 0$, $K_{n+1} = (n+1)^{-1}c\Delta/(2g)$ depending on c. $\bar{\phi}_1(y)$ reads then

$$\bar{\phi}_1(y) = cy^n - \Delta\sum_{m=n+1}^{\infty} K_m\frac{y^m}{m-n}. \tag{21}$$

Because c multiplies both $\bar{\phi}_1$ and $\bar{\phi}_2$, it may be set to $2g(n+1)/\Delta$. The solution will have parity $\sigma \in \{1, -1\}$ and be analytic in all of $\mathbb{C}$, if the G-function

$$G_\sigma^{(n)}(g,\Delta) = -\sigma\frac{2(n+1)}{\Delta} + \sum_{m=1}^{\infty} K_{n+m}\left(1 + \sigma\frac{\Delta}{m}\right)g^{m-1} \tag{22}$$

vanishes for parameters g, Δ. One sees immediately that $G_+^{(n)} = G_-^{(n)} = 0$ entails $\bar{\phi}_1(z+g) = \bar{\phi}_2(z+g) \equiv 0$, thus this state is non-degenerate if it exists. States of this type comprise the non-degenerate exceptional spectrum [42] and are characterized by a lifting of the pole of $G_+(x)$ (resp. $G_-(x)$) at $x = n$ for special values of g, Δ, satisfying $G_\pm^{(n)}(g,\Delta) = 0$. The exceptional G-functions in Equation (22) are given in terms of absolutely convergent series expansions as the regular G-functions in Equation (10). The other possible Frobenius solution at $y = 0$, $\bar{\phi}_2(y) = \sum_{m=0}^{\infty} K_m y^m$, leads to

$$\bar{\phi}_1(y) = cy^n - \Delta\sum_{m\neq n}^{\infty} K_m\frac{y^m}{m-n} - \Delta K_n y^n \ln(y), \tag{23}$$

where the K_m for $m \leq n$ are determined with the same recurrence as above and initial conditions $K_{-1} = 0$, $K_0 = 1$, which fixes the overall factor of the wavefunction. This solution is only independent from the first and admissible if $n \geq 1$ and $K_n(n) = 0$. If so, the K_m for $m \geq n+1$ are computed recursively via Equation (11) with initial conditions $K_n = 0$, $K_{n+1} = (n+1)^{-1}[c\Delta/(2g) - K_{n-1}]$. Parity symmetry determines now the constant $c(\sigma)$,

$$\sum_{m\neq n}^{\infty} K_m\left(1 + \sigma\frac{\Delta}{m-n}\right)g^m - c(\sigma)g^n = 0. \tag{24}$$

Equation (24) imposes no additional constraint on g, Δ besides $K_n(n) = 0$, which is therefore sufficient for the presence of a doubly degenerate solution with $x = n$. Because $n \geq 1$, this type of degenerate solution cannot occur for $x = 0$, whereas non-degenerate solutions with $x = 0$ are possible.

For the choice $c = 2gK_{n-1}/\Delta$, one of the degenerate solutions reads

$$\begin{aligned} \bar{\phi}_2(y) &= \sum_{m=0}^{n-1} K_m y^n, \\ \bar{\phi}_1(y) &= \Delta \sum_{m=0}^{n-1} K_m \frac{y^m}{n-m} + \frac{2gK_{n-1}}{\Delta} y^n. \end{aligned} \tag{25}$$

The $\bar{\phi}_j(y)$ are polynomials in y, therefore Equation (25) is a quasi-exact solution with polynomial wave function, apart from the factor e^{-gz} multiplying $\bar{\phi}_{1,2}$ in $\phi_{1,2}$. This quasi-exact solution is not a parity eigenstate but a linear combination of them. The parity eigenstates are in turn a linear combination of Equation (25) and states having the form of non-degenerate exceptional solutions. It is clear that the possibility of quasi-exact solutions in the QRM depends on the fact that the coefficients of the Frobenius solutions are determined by a three-term recurrence relation (Equation (11)). Otherwise, the single free integration constant c would not suffice to break off the series expansions for $\bar{\phi}_{1,2}$ at finite order. This is the reason a quasi-exact spectrum does not exist in the isotropic Dicke model [27] but is possible in the anisotropic Dicke models, where more parameters can be adjusted to eliminate the higher orders in expansions given by recurrence relations with more than three terms [43].

5. Conclusions

The quantum Rabi model is the most simple theoretical description of the interaction between light and matter at strong coupling. Despite its simplicity, its spectrum displays many interesting and unusual features such as two-fold degeneracies confined to baselines, the almost equally spaced distribution of eigenvalues along the real axis and the quasi-exact spectrum. All these peculiarities can be traced back to the integrability of the quantum Rabi model, i.e. the fact that the Hilbert space of the discrete degree of freedom is two-dimensional and therefore equals the number of irreducible representations of its symmetry group, $\mathbb{Z}_2$. This symmetry also causes the qualitative deviations of the Rabi spectrum from the Jaynes–Cummings spectrum, although they coincide almost perfectly for small coupling. The Jaynes–Cummings model possesses a much larger continuous $U(1)$-symmetry and therefore many more level crossings in the spectral graph. Any approximation of the QRM which employs a kind of rotating-wave approximation introduces automatically this $U(1)$-symmetry and the concomitant unphysical level crossings, even if they do not occur in certain parameter ranges to which these approximations are thus confined.

Funding: This research received no external funding.

Acknowledgments: I wish to thank Michael Dzierzawa for providing Figure 4.

Conflicts of Interest: The author declares no conflict of interest.

References

1. Cohen-Tannoudji, C.; Dupont-Roc, J.; Grynberg, G. *Atom-Photon Interactions: Basic Processes and Applications*; Wiley-VCH: Hamburg, Germany, 1992.
2. Rabi, I.I. On the Process of Space Quantization. *Phys. Rev.* **1936**, *49*, 324–328. [CrossRef]
3. Rabi, I.I. Space Quantization in a Gyrating Magnetic Field. *Phys. Rev.* **1937**, *51*, 652–654. [CrossRef]

4. Jaynes, E.T.; Cummings, F.W. Comparison of quantum and semiclassical radiation theories with application to the beam maser. *Proc. IEEE* **1963**, *51*, 89–109. [CrossRef]
5. Allen, L.; Eberly, J.H. *Optical Resonance and Two-Level Atoms*; Wiley: New York, NY, USA, 1987.
6. Amico, L.; Frahm, H.; Osterloh, A.; Ribeiro, G.A.P. Integrable spin–boson models descending from rational six-vertex models. *Nucl. Phys. B* **2007**, *787*, 283–300. [CrossRef]
7. Braak, D. Integrability of the Rabi Model. *Phys. Rev. Lett.* **2011**, *107*, 100401. [CrossRef]
8. Thompson, R.J.; Rempe, G.; Kimble, H.J. Observation of normal-mode splitting for an atom in an optical cavity. *Phys. Rev. Lett.* **1992**, *68*, 1132–1135. [CrossRef] [PubMed]
9. Symonds, C.; Bonnand, C.; Plenet, J.C.; Bréhier, A.; Parashkov, R.; Lauret, J.S.; Deleporte, E.; Bellessa, J. Particularities of surface plasmon–exciton strong coupling with large Rabi splitting. *New J. Phys.* **2008**, *10*, 065017. [CrossRef]
10. Forn-Díaz, P.; Lamata, L.; Rico, E.; Kono, J.; Solano, E. Ultrastrong coupling regimes of light-matter interaction. *Rev. Mod. Phys.* **2019**, *91*, 025005. [CrossRef]
11. Casanova, J.; Romero, G.; Lizuain, I.; García-Ripoll, J.J.; Solano, E. Deep Strong Coupling Regime of the Jaynes-Cummings Model. *Phys. Rev. Lett.* **2010**, *105*, 263603. [CrossRef]
12. Yoshihara, F.; Fuse, T.; Ashhab, S.; Kakuyanagi, K.; Saito, S.; Semba, K. Superconducting qubit-oscillator circuit beyond the ultrastrong-coupling regime. *Nat. Phys.* **2017**, *13*, 44–47. [CrossRef]
13. Niemczyk, T.; Deppe, F.; Huebl, H.; Menzel, E.P.; Hocke, F.; Schwarz, M.J.; Garcia-Ripoll, J.J.; Zueco, D.; Hümmer, T.; Solano, E.; et al Circuit quantum electrodynamics in the ultrastrong-coupling regime. *Nat. Phys.* **2010**, *6*, 772–776. [CrossRef]
14. Bloch, F.; Siegert, A. Magnetic Resonance for Nonrotating Fields. *Phys. Rev.* **1940**, *57*, 522–527. [CrossRef]
15. Klimov, A.B.; Chumakov, S.M. *A Group-Theoretical Approach to Quantum Optics: Models of Atom-Field Interactions*; John Wiley & Sons: Weinheim, Germany, 2009.
16. Nielsen, M.A.; Chuang, I.L.; Chuang, I.L. *Quantum Computation and Quantum Information*; Cambridge University Press: Cambridge, UK, 2000.
17. Schuster, D.I.; Wallraff, A.; Blais, A.; Frunzio, L.; Huang, R.S.; Majer, J.; Girvin, S.M.; Schoelkopf, R.J. ac Stark Shift and Dephasing of a Superconducting Qubit Strongly Coupled to a Cavity Field. *Phys. Rev. Lett.* **2005**, *94*, 123602. [CrossRef] [PubMed]
18. Felicetti, S.; Sanz, M.; Lamata, L.; Romero, G.; Johansson, G.; Delsing, P.; Solano, E. Dynamical Casimir Effect Entangles Artificial Atoms. *Phys. Rev. Lett.* **2014**, *113*, 093602. [CrossRef] [PubMed]
19. Córcoles, A.D.; Magesan, E.; Srinivasan, S.J.; Cross, A.W.; Steffen, M.; Gambetta, J.M.; Chow, J.M. Demonstration of a quantum error detection code using a square lattice of four superconducting qubits. *Nat. Commun.* **2015**, *6*, 6979. [CrossRef] [PubMed]
20. Ciuti, C.; Bastard, G.; Carusotto, I. Quantum vacuum properties of the intersubband cavity polariton field. *Phys. Rev. B* **2005**, *72*, 115303. [CrossRef]
21. De Liberato, S. Light-Matter Decoupling in the Deep Strong Coupling Regime: The Breakdown of the Purcell Effect. *Phys. Rev. Lett.* **2014**, *112*, 016401. [CrossRef] [PubMed]
22. Feranchuk, I.D.; Komarov, L.I.; Ulyanenkov, A.P. Two-level system in a one-mode quantum field: Numerical solution on the basis of the operator method. *J. Phys. A Math. Gen.* **1996**, *29*, 4035–4047. [CrossRef]
23. Irish, E.K. Generalized Rotating-Wave Approximation for Arbitrarily Large Coupling. *Phys. Rev. Lett.* **2007**, *99*, 173601. [CrossRef]
24. Gan, C.J.; Zheng, H. Dynamics of a two-level system coupled to a quantum oscillator: Transformed rotating-wave approximation. *Eur. Phys. J. D* **2010**, *59*, 473–478. [CrossRef]
25. Judd, B.R. Exact solutions to a class of Jahn-Teller systems. *J. Phys. C Solid State Phys.* **1979**, *12*, 1685–1692. [CrossRef]
26. Arnol'd, V.I. *Mathematical Methods of Classical Mechanics*, 2nd ed.; Graduate Texts in Mathematics; Springer: New York, NY, USA, 1989.
27. Braak, D. Solution of the Dicke model forN= 3. *J. Phys. B At. Mol. Opt. Phys.* **2013**, *46*, 224007. [CrossRef]

28. Tavis, M.; Cummings, F.W. Exact Solution for an N-Molecule—Radiation-Field Hamiltonian. *Phys. Rev.* **1968**, *170*, 379–384. [CrossRef]
29. Miller, W.; Post, S.; Winternitz, P. Classical and quantum superintegrability with applications. *J. Phys. A Math. Theor.* **2013**, *46*, 423001. [CrossRef]
30. Caux, J.S.; Mossel, J. Remarks on the notion of quantum integrability. *J. Stat. Mech.* **2011**, *2011*, P02023. [CrossRef]
31. Eckle, H.P. *Models of Quantum Matter: A First Course on Integrability and the Bethe Ansatz*; Oxford University Press: Oxford, UK, 2019.
32. Berry, M.V.; Tabor, M. Level clustering in the regular spectrum. *Proc. R. Soc. Lond. Math. Phys. Sci.* **1977**, *356*, 375–394. [CrossRef]
33. Batchelor, M.T.; Zhou, H.Q. Integrability versus exact solvability in the quantum Rabi and Dicke models. *Phys. Rev. A* **2015**, *91*, 053808. [CrossRef]
34. Kuś, M. Statistical Properties of the Spectrum of the Two-Level System. *Phys. Rev. Lett.* **1985**, *54*, 1343–1345. [CrossRef] [PubMed]
35. Mehta, M.L. *Random Matrices*; Elsevier: London, UK, 2004.
36. Slavyanov, S.Y.; Lay, W. *Special Functions: A Unified Theory Based on Singularities*; Oxford University Press: Oxford, UK, 2000.
37. Zhong, H.; Xie, Q.; Batchelor, M.T.; Lee, C. Analytical eigenstates for the quantum Rabi model. *J. Phys. A Math. Theor.* **2013**, *46*, 415302. [CrossRef]
38. Sugiyama, S. Spectral Zeta Functions for the Quantum Rabi Models. *Nagoya Math. J.* **2016**, 1–47. [CrossRef]
39. Felicetti, S.; Rossatto, D.Z.; Rico, E.; Solano, E.; Forn-Díaz, P. Two-photon quantum Rabi model with superconducting circuits. *Phys. Rev. A* **2018**, *97*. [CrossRef]
40. Duan, L.; Xie, Y.F.; Braak, D.; Chen, Q.H. Two-photon Rabi model: Analytic solutions and spectral collapse. *J. Phys. A Math. Theor.* **2016**, *49*, 464002. [CrossRef]
41. Ince, E.L. *Ordinary Differential Equations*; Dover: Gand County, UK, 2012.
42. Maciejewski, A.J.; Przybylska, M.; Stachowiak, T. Full spectrum of the Rabi model. *Phys. Lett. A* **2014**, *378*, 16–20. [CrossRef]
43. Peng, J.; Zheng, C.; Guo, G.; Guo, X.; Zhang, X.; Deng, C.; Ju, G.; Ren, Z.; Lamata, L.; Solano, E. Dark-like states for the multi-qubit and multi-photon Rabi models. *J. Phys. A Math. Theor.* **2017**, *50*, 174003. [CrossRef]

symmetry

MDPI

Article

Behavior of Floquet Topological Quantum States in Optically Driven Semiconductors

Andreas Lubatsch [1,†] **and Regine Frank** [2,3,*,†]

1 University of Applied Sciences Nürnberg Georg Simon Ohm, Keßlerplatz 12, 90489 Nürnberg, Germany; lubatsch@th.physik.uni-bonn.de
2 Bell Labs, 600 Mountain Avenue, Murray Hill, NJ 07974-0636, USA
3 Serin Physics Laboratory, Department of Physics and Astronomy, Rutgers University, 136 Frelinghuysen Road, Piscataway, NJ 08854-8019, USA
* Correspondence: regine.frank@rutgers.edu or regine.frank@googlemail.com
† These authors contributed equally to this work.

Received: 22 July 2019; Accepted: 18 September 2019; Published: 4 October 2019

Abstract: Spatially uniform optical excitations can induce Floquet topological band structures within insulators which can develop similar or equal characteristics as are known from three-dimensional topological insulators. We derive in this article theoretically the development of Floquet topological quantum states for electromagnetically driven semiconductor bulk matter and we present results for the lifetime of these states and their occupation in the non-equilibrium. The direct physical impact of the mathematical precision of the Floquet-Keldysh theory is evident when we solve the driven system of a generalized Hubbard model with our framework of dynamical mean field theory (DMFT) in the non-equilibrium for a case of ZnO. The physical consequences of the topological non-equilibrium effects in our results for correlated systems are explained with their impact on optoelectronic applications.

Keywords: topological excitations; Floquet; dynamical mean field theory; non-equilibrium; stark-effect; semiconductors

PACS: 71.10.-w theories and models of many-electron systems; 42.50.Hz strong-field excitation of optical transitions in quantum systems; multi-photon processes; dynamic Stark shift; 74.40+ Fluctuations; 03.75.Lm Tunneling, Josephson effect, Bose-Einstein condensates in periodic potentials, solitons, vortices, and topological excitations; 72.20.Ht high-field and nonlinear effects; 89.75.-k complex systems

---

1. Introduction

Topological phases of matter [1–3] have captured our fascination over the past decades, revealing properties in the sense of robust edge modes and exotic non-Abelian excitations [4,5]. Potential applications of periodically driven quantum systems [6] are conceivable in the subjects of semiconductor spintronics [7] up to topological quantum computation [8] as well as topological lasers [9,10] in optics and random lasers [11]. Already topological insulators in solid-state devices such as HgTe/CdTe quantum wells [12,13], as well as topological Dirac insulators such as Bi_2Te_3 and Bi_2Sn_3 [14–16] were groundbreaking discoveries in the search for the unique properties of topological phases and their technological applications.

In non-equilibrium systems, it has been shown that time-periodic perturbations can induce topological properties in conventional insulators [17–20] which are trivial in equilibrium otherwise. Floquet topological insulators include a very broad range of physical solid state and atomic realizations, driven at resonance or off-resonance. These systems can display metallic conduction, which is enabled by quasi-stationary states at the edges [17,21,22]. Their band structure may have the form of a

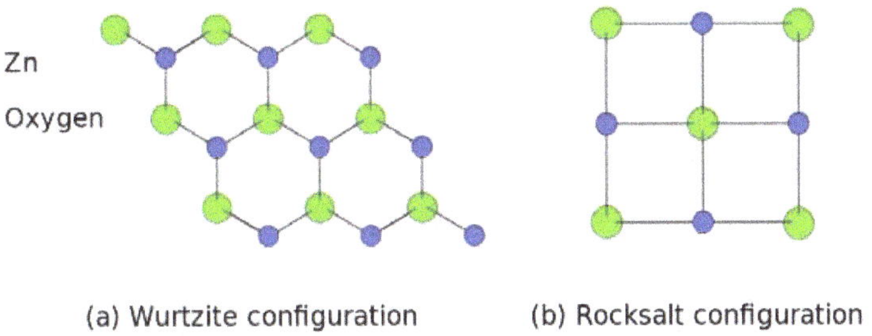

Figure 1. ZnO structure (ab-plane). (**a**) non-centrosymmetric, hexagonal, wurtzite configuration; (**b**) centrosymmetric, cubic, rocksalt configuration (Rochelle salt) [30–32]. The rocksalt configuration is distinguished by a tunable gap from 1.8 eV up to 6.1 eV, a gap value of 2.45 eV is typical for the monocrystal rocksalt configuration without oxygen vacancies [33,34]. As such, the rocksalt configuration could be suited for higher harmonics generation under non-equilibrium topological excitation [35,36].

Dirac cone in three-dimensional systems [23,24], and Floquet Majorana fermions [25] have been conceptionally developed. Graphene and Floquet fractional Chern insulators have been recently investigated [26–28].

In this article, we show that Floquet topological quantum states can evolve in correlated electronic systems of driven semi-conductors in the non-equilibrium. We investigate ZnO bulk matter in the centrosymmetric, cubic rocksalt configuration, see Figure 1. The non-equilibrium is in this sense defined by the intense external electromagnetic driving field, which induces topologically dressed electronic states and the evolution of dynamical gaps, see Figure 2. These procedures are expected to be observable in pump-probe experiments on time scales below the thermalization time. We show that the expansion into Floquet modes [29], see Figure 3, is leading to results of direct physical impact in the sense of modeling the coupling of a classical electromagnetic external driving field to the correlated quantum many body system. Our results derived by Dynamical Mean Field Theory (DMFT) in the non-equilibrium provide novel insights in topologically induced phase transitions of driven otherwise conventional three-dimensional semiconductor bulk matter and insulators.

2. Quantum Many Body Theory for Correlated Electrons in the Non-Equilibrium

We consider in this work the wide gap semiconductor bulk to be driven by a strong periodic-in-time external field in the optical range which yields higher-order photon absorption processes. The electronic dynamics of the photo-excitation processes, see Figure 2, is theoretically modelled by a generalized, driven, Hubbard Hamiltonian, see Equation (1). The system is solved with a Keldysh formalism including the electron-photon interaction in the sense of the coupling of the classical electromagnetic field to the electronic dipole and thus to the electronic hopping. This yields an additional kinetic contribution. We solve the system by the implementation of a dynamical mean field theory (DMFT), see Figure 4, with a generalized iterative perturbation theory solver (IPT), see Figure 5. The full interacting Hamiltonian, Equation (1), is introduced as follows:

$$\begin{aligned} H \;=\; & \sum_{i,\sigma} \varepsilon_i c^{\dagger}_{i,\sigma} c_{i,\sigma} + \frac{U}{2} \sum_{i,\sigma} c^{\dagger}_{i,\sigma} c_{i,\sigma} c^{\dagger}_{i,-\sigma} c_{i,-\sigma} \\ & - t \sum_{\langle ij \rangle,\sigma} c^{\dagger}_{i,\sigma} c_{j,\sigma} \\ & + i\vec{d} \cdot \vec{E}_0 \cos(\Omega_L \tau) \sum_{<ij>,\sigma} \left(c^{\dagger}_{i,\sigma} c_{j,\sigma} - c^{\dagger}_{j,\sigma} c_{i,\sigma} \right). \end{aligned} \tag{1}$$

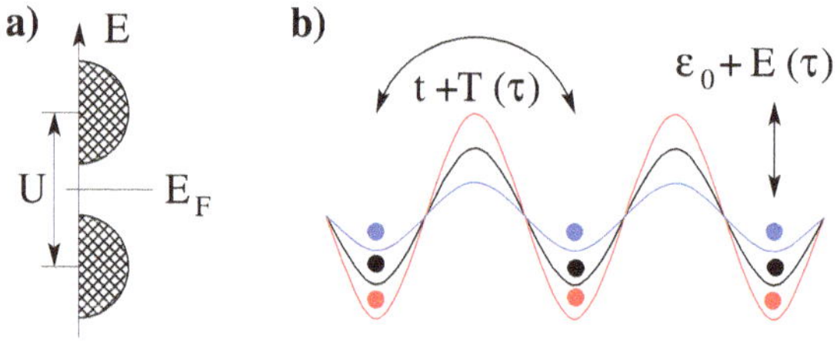

Figure 2. Insulator to metal transition caused by photo-excitation. (**a**) schematic split of energy bands due to the local Coulomb interaction U. The gap is determined symmetrically to the Fermi edge E_F; (**b**) the periodic in time driving yields an additional hopping contribution $T(\tau)$ of electrons on the lattice (black) and the renormalization of the local potential, $E(\tau)$, as a quasi-energy. Colors of the lattice potential represent the external driving in time.

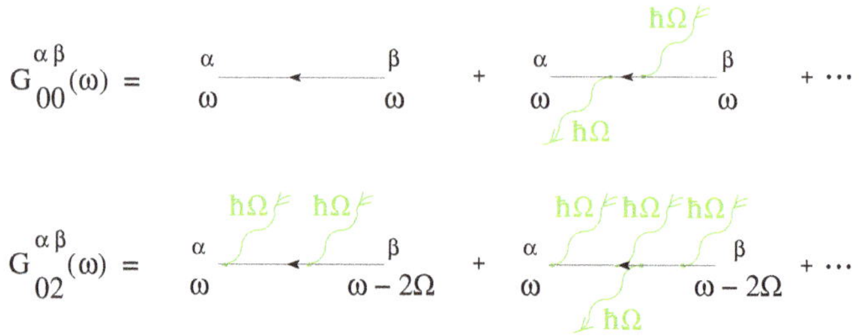

Figure 3. Schematic representation of the Floquet Green's function and the Floquet matrix in terms of absorption and emission of external energy quanta $\hbar\Omega$. $G^{\alpha\beta}_{00}(\omega)$ represents the sum of all balanced contributions; $G^{\alpha\beta}_{02}(\omega)$ describes the net absorption of two photons. α, β are the Keldysh indices.

In our notation, see Equation (1), $c^\dagger, (c)$ are the creator (annihilator) of an electron. The subscripts i, j indicate the site, $\langle i, j\rangle$ implies the sum over nearest neighboring sites.

The term $\frac{U}{2}\sum_{i,\sigma} c^\dagger_{i,\sigma}c_{i,\sigma}c^\dagger_{i,-\sigma}c_{i,-\sigma}$ results from the repulsive onsite Coulomb interaction U between electrons with opposite spins. The third term $-t\sum_{\langle ij\rangle,\sigma}c^\dagger_{i,\sigma}c_{j,\sigma}$ describes the standard hopping processes of electrons with the amplitude t between nearest neighboring sites. Those contributions form the standard Hubbard model, which is generalized for our purposes in what follows. The first term $\sum_{i,\sigma}\varepsilon_i c^\dagger_{i,\sigma}c_{i,\sigma}$ generalizes the Hubbard model with respect to the onsite energy, see Figure 2. The electronic on-site energy is noted as ε_i. The external time-dependent electromagnetic driving is described in terms of the field $\vec{E}_0$ with laser frequency Ω_L, τ, which couples to the electronic dipole $\hat{d}$ with strength $|\mathbf{d}|$. The expression $i\vec{d}\cdot\vec{E}_0\cos(\Omega_L\tau)\sum_{<ij>,\sigma}\left(c^\dagger_{i,\sigma}c_{j,\sigma} - c^\dagger_{j,\sigma}c_{i,\sigma}\right)$ describes the renormalization of the standard electronic hopping processes, as one possible contribution $T(\tau)$ in Figure 2, due to external influences.

2.1. Floquet States: Coupling of a Classical Driving Field to a Quantum Dynamical System

By introducing the explicit time dependency of the external field, we solve the generalized Hubbard Hamiltonian, see Equation (1). It yields Green's functions which depend on two separate time arguments which are Fourier transformed to frequency coordinates. These frequencies are chosen as the relative and the center-of-mass frequency [38,39] and we introduce an expansion into Floquet modes

$$\begin{aligned} G^{\alpha\beta}_{mn}(\omega) &= \int d\tau_1^\alpha d\tau_2^\beta\, e^{-i\Omega_L(m\tau_1^\alpha - n\tau_2^\beta)} e^{i\omega(\tau_1^\alpha - \tau_2^\beta)} G(\tau_1^\alpha, \tau_2^\beta) \\ &\equiv G^{\alpha\beta}(\omega - m\Omega_L, \omega - n\Omega_L). \end{aligned} \tag{2}$$

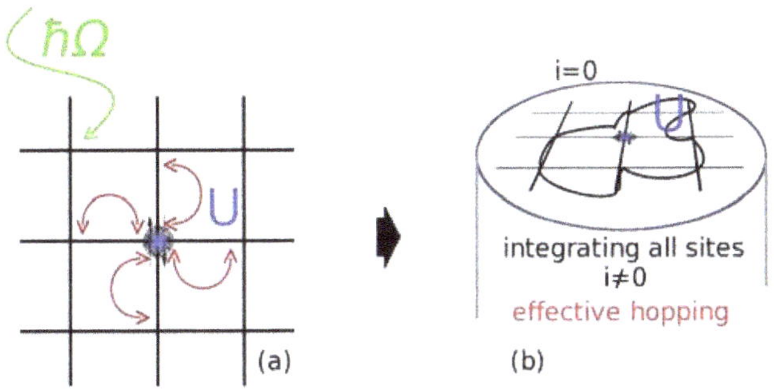

Figure 4. Schematic representation of non-equilibrium dynamical mean field theory. (**a**) the semiconductor behaves in the here considered regime as an insulator: Optical excitations by an external electromagnetic field with the energy $\hbar\Omega$ yield additional hopping processes. These processes are mapped onto the interaction with the single site on the background of the surrounding lattice bath in addition to the regular kinetic processes and in addition to on-site Coulomb repulsion; (**b**) DMFT idea: The integration over all lattice sites leads to an effective theory including non-equilibrium excitations. The bath consists of all single sites and the approach is thus self-consistent. The driven electronic system may in principal couple to a surface-resonance or an edge state. The coupling to these states can be enhanced by the external excitation.

In general, Floquet [29] states are analogues to Bloch states. Whereas Bloch states are due to the periodicity of the potential in space, the spatial topology, the Floquet states represent the temporal topology in the sense of the temporal periodicity [35,38–46]. The Floquet expansion is introduced in Figure 3 as a direct graphic representation of what is described in Equation (2). The Floquet modes are labelled by the indices (m, n), whereas (α, β) refer to the branch of the Keldysh contour $(\pm)$ and the respective time argument. The physical consequence of the Floquet expansion, however, is noteworthy, since it can be understood as the quantized absorption and emission of energy $\hbar\Omega_L$ by the driven quantum many body system out of and into the classical external driving field.

In the case of uncorrelated electrons, $U = 0$, the Hamiltonian can be solved analytically and the retarded component of the Green's function $G_{mn}(k, \omega)$ reads

$$G^R_{mn}(k,\omega) = \sum_{\rho} \frac{J_{\rho-m}\left(A_0\tilde{\epsilon}_k\right) J_{\rho-n}\left(A_0\tilde{\epsilon}_k\right)}{\omega - \rho\Omega_L - \epsilon_k + i0^+}. \tag{3}$$

Here, $\tilde{\epsilon}_k$ is the dispersion relation induced by the external driving field. $\tilde{\epsilon}_k$ is to be distinguished from the lattice dispersion ϵ. J_n are the cylindrical Bessel functions of integer order, $A_0 = \vec{d} \cdot \vec{E}_0$, Ω_L is the external laser frequency. The retarded Green's function for the optically excited band electron is eventually given by

$$G^R_{\mathrm{Lb}}(k,\omega) = \sum_{m,n} G^R_{mn}(k,\omega). \tag{4}$$

2.2. *Dynamical Mean Field Theory in the Non-Equilibrium*

The generalized Hubbard model for the correlated system, $U \neq 0$, in the non-equilibrium, Equation (1), is numerically solved by a single-site Dynamical Mean Field Theory (DMFT) [37,46–59]. The expansion into Floquet modes with the proper Keldysh description models the external time dependent classical driving field, see Section 2.1, and couples it to the quantum many body system. We numerically solve the Floquet-Keldysh DMFT [37,46] with a second order iterative perturbation theory (IPT), where the the local self-energy $\Sigma^{\alpha\beta}$ is derived by four bubble diagrams; see Figure 5.

The Green's function for the interaction of the laser with the band electron $G^R_{\mathrm{Lb}}(k,\omega)$, Equation (4), is characterized by the wave vector k, where k describes the periodicity of the lattice. It depends on the electronic frequency ω and the external driving frequency Ω_L, see Equation (2), captured in the Floquet indices (m,n). The DMFT self-consistency relation assumes the form of a matrix equation of non-equilibrium Green's functions, which is of dimension 2×2 in regular Keldysh space and of dimension $n\times n$ in Floquet space. The numerical algorithm is efficient and stable also for all values of the Coulomb interaction U.

In previous work [37,46,56], we considered an additional kinetic energy contribution due to a lattice vibration. Here, we take into account a coupling of the microscopic electronic dipole moment to an external electromagnetic field [38,39] for a correlated system. We introduce the quantum-mechanical expression for the electronic dipole operator $\hat{d}$, see the last term r.h.s. Equation (1), and this coupling reads as $i\vec{d}\cdot\vec{E}_0\cos(\Omega_L\tau)\sum_{<ij>,\sigma}\left(c^{\dagger}_{i,\sigma}c_{j,\sigma}-c^{\dagger}_{j,\sigma}c_{i,\sigma}\right)$. This kinetic contribution is conceptually different from the generic kinetic hopping of the third term of Equation (1). The coupling $\hat{d}\cdot\vec{E}_0\cos(\Omega_L\tau)$ generates a factor Ω_L that cancels the $1/\Omega_L$ in the renormalized cylindrical Bessel function in Equation (7) of Ref. [37] in the Coulomb gauge, $\vec{E}(\tau) = -\frac{\partial}{\partial\tau}\vec{A}(\tau)$ that is written in Fourier space as $\vec{E}(\Omega_L) = i\Omega_L\cdot\vec{A}(\Omega_L)$. The Floquet sum, which is a consistency check, is discussed in Section 3.3.

It has been shown by Ref. [49] that the coupling of an electromagnetic field modulation to the onsite electronic density $n_i = c^{\dagger}_{i,\sigma}c_{i,\sigma}$ in the unlimited three-dimensional translationally invariant system alone can be gauged away. This type of coupling can be absorbed in an overall shift of the local potential while no additional dispersion is reflecting any additional functional dynamics of the system. Therefore, such a system [26,60] will not show any topological effects as a topological insulator or a Chern insulator. In contrast, the coupling of the external electromagnetic field modulation to the dipole moment of the charges, and thus to the hopping term, see Equation (1), as a kinetic energy of the fermions, cannot be gauged away and is causing the development of topological states in the three-dimensional unlimited systems. A boundary as such is no necessary requirement. Line 3 of Equation (1) formally represents the electromagnetically induced kinetic contribution

$$i\vec{d}\cdot\vec{E}_0\cos(\Omega_L\tau)\sum_{<ij>,\sigma}\left(c^{\dagger}_{i,\sigma}c_{j,\sigma}-c^{\dagger}_{j,\sigma}c_{i,\sigma}\right) = e\sum_{\vec{r}}\hat{j}_{ind}(\vec{r})\cdot\vec{A}(\vec{r},\tau), \tag{5}$$

which is the kinetic contribution of the photo-induced charge current in-space dependent with $\vec{r}$

$$\vec{j}_{ind}(\vec{r})_{\delta} = -\frac{t}{i}\sum_{\sigma}(c^{\dagger}_{\vec{r},\sigma}c_{\vec{r}+\delta,\sigma} - c^{\dagger}_{\vec{r}+\delta,\sigma}c_{\vec{r},\sigma}). \tag{6}$$

The temporal modulation of the classical external electrical field in the (111) direction always causes a temporally modulated magnetic field contribution $\vec{B}(\vec{r},\tau) = \nabla\times\vec{A}(\vec{r},\tau)$ with $\vec{B}(\vec{r},\Omega_L)$ in Fourier space, as a consequence of Maxwell's equations. In the following, we derive the non-equilibrium local density of states (LDOS) which comes along with the dynamical life-time of non-equilibrium states as an inverse of the imaginary part of the self-energy $\tau\sim 1/\Im\Sigma^R$. A time reversal procedure induced by an external field will never be able to revise the non-equilibrium effect. The photon-electron coupling and thus the absorption will be modified and overall profoundly differing material characteristics are created. Conductivity and polarization of excited matter in the non-equilibrium are preventing any time-reversal processes in the sense of closing the Floquet fan again in this regime. The initial electromagnetic field thus causes a break of the time-reversal symmetry, and the current leads to the acquisition of a non-zero Berry flux. A Wannier-Stark type ladder [61] is created, which can be characterized by its Berry phase [62] as a Chern or a winding number or the $\mathbb{Z}_2$ invariants in three dimensions respectively [63].

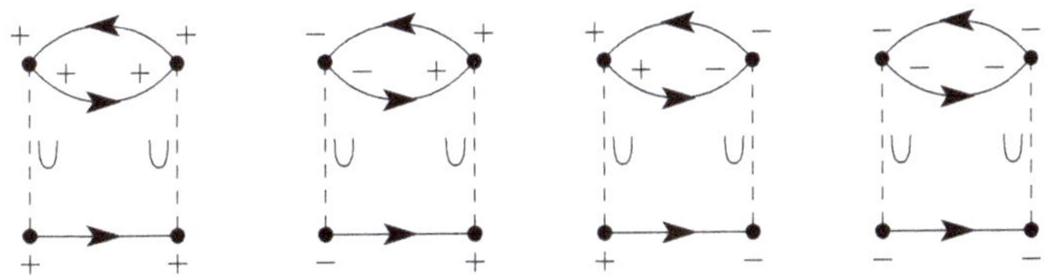

Figure 5. Local self-energy $\Sigma^{\alpha\beta}$ within the iterated perturbation theory (IPT). The IPT as a second order diagrammatic solver with respect to the electron electron interaction U is here generalized to non-equilibrium, $\pm$ indicates the branch of the Keldysh contour. The solid lines represent the bath in the sense of the Weiss-field $\mathcal{G}^{\alpha\beta}$; see Ref. [37].

3. Floquet Spectra of Driven Semiconductors

From the numerically computed components of the Green's function, we define [37] the local density of states (LDOS), $N(\omega, \Omega_L)$, where momentum is integrated out and Floquet indices are summed

$$N(\omega, \Omega_L) = -\frac{1}{\pi} \sum_{mn} \int \mathrm{d}^3k \mathrm{Im}\, G^R_{mn}(\mathbf{k}, \omega, \Omega_L). \tag{7}$$

In combination with the lifetime as the inverse of the imaginary part of the self-energy, $\tau \sim 1/\Im\Sigma^R$, and the non-equilibrium distribution function

$$F^{neq}(\omega, \Omega_L) = \frac{1}{2}\left(1 + \frac{1}{2i} \frac{\sum_m G^{Keld}_{0m}(\omega, \Omega_L)}{\sum_n \mathrm{Im} G^A_{0n}(\omega, \Omega_L)}\right), \tag{8}$$

the local density of states $N(\omega, \Omega_L)$ can be experimentally determined as the compelling band structure of the non-equilibrium system.

We show results for optically excited semiconductor bulk, with a band gap in the equilibrium of 2.45 eV and typical parameters for ZnO. ZnO in either configuration [33,34,64–66] is a very promising material for the construction of micro-lasers, quantum wells and optical components. In certain geometries and in connection to other topological insulators, it is already used for the engineering of ultrafast switches. ZnO, see Figure 1, is broadly investigated in the non-centro-symmetric wurtzite configuration and very recently in the centro-symmetric rocksalt configuration [31,32]. Its bandgap is estimated to be of 1.8 eV up to 6.1 eV depending on various factors as the pressure during the fabrication process. In either crystal configuration, the production of second or higher order harmonics under intense external excitations [67] is searched. It is of high interest for novel types of lasers.

3.1. Development and Lifetimes of Floquet Topological Quantum States in the Non-Equilibrium

In Figure 6, we investigate a wide gap semi-conductor band structure, and the band gap in equilibrium is assumed to be 2.45 eV. The semiconductor bulk shall be exposed to an external periodic-in-time driving field. The system is so far considered as pure bulk, so we are investigating Floquet topological effects in the non-equilibrium without any other geometrical influence. The excitation intensity in the results of Figure 6a is considered to be 5.0 MW/cm^2 and 10.0 MW/cm^2 in Figure 6b. DMFT as a solver for correlated and strongly correlated electronics as such is a spatially independent method. It is designed to derive bulk effects, whereas all k-dependencies have been integrated as the fundamental methodology. Therefore, we are not analyzing the k-resolved information of the Brillouin zone. As long as no artificial coarse graining with a novel length scale in the sense of finite elements or finite volumes is included, DMFT results in one, two and three dimensions are independent of any spatial information. In fact, however, the energy dependent LDOS profoundly changes with a varying excitation frequency and with a varying excitation intensity as well, which gives evidence that also the underlying k-dependent band structure is topologically

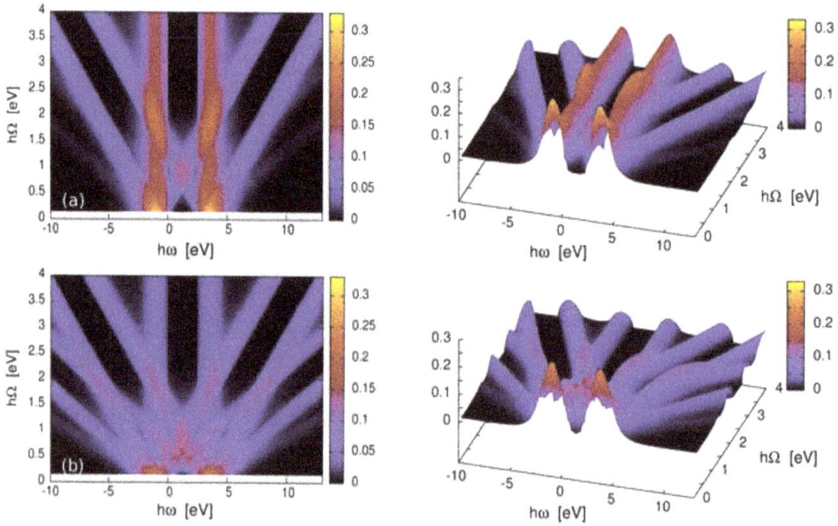

Figure 6. Floquet topological quantum states of the semiconductor bulk in the non-equilibrium. (**a**) the evolution of the LDOS in the non-equilibrium for varying excitation laser frequencies Ω_L up to $\Omega_L = 4.0$ eV is shown. The excitation intensity 5.0 MW/cm^2 is constant. The bandgap of ZnO rocksalt in equilibrium is 2.45 eV, see Figure 2, the gap is vanishing with the increase of the driving frequency and dressed states emerge as a consequence of the non-equilibrium AC-Stark effect [82,83]. The split bands are superposed by a doublet of Floquet fans which intersect. The formation of topological subgaps, see e.g., at $\hbar\Omega_L = 0.9$ eV occurs; (**b**) the evolution of the LDOS for the excitation intensity of 10.0 MW/cm^2 is shown. Spectral weight is shifted to a multitude of higher order Floquet-bands, while the original split band characteristics almost vanishes apart from the near-gap band edges. A variety of Floquet gaps is formed. At any crossing point, topologically induced transitions are possible, and the generation of higher harmonics can be enhanced. Panels on the right display the topology of the LDOS. The subgaps are very pronounced and the intersection of bands is visible as an increase of the LDOS which can be measurable in a pump-probe experiment. For a detailed discussion, please see Section 3.

modulated. A non-trivial topological structure of the Hilbert space is generated by external excitations even though our system in equilibrium is fully periodic in space and time. The time dependent external electrical field generates a temporally modulated magnetic field which results in a dynamical Wannier-Stark effect and the generation of Floquet states. Floquet states are the temporal analogue to Bloch states, and thus the argumentation by Zak [61] in principle applies for the generation of the Berry phase γ_m, since the solid is exposed to an externally modulated electromagnetic potential [62,68–70]. The Floquet quasi-energies, see Figure 3, are labeled by the Floquet modes in dependency to the external excitation frequency, and to the external excitation amplitude. The topological invariants, the Chern number as a sum over all occupied bands $n = \sum_{m=1}^{\nu} n_m \neq 0$ and the $\mathbb{Z}_2$ invariants include the Berry flux $n_m = 1/2\pi \int d^2\vec{k}(\nabla \times \gamma_m)$. The winding number is also consistently associated with the argument of collecting a non-zero Berry flux. We consider both regimes, where the driving frequency is smaller than the width of the semiconductor gap in equilibrium and also where it is larger and a very pronounced topology of states is generated. While the system is excited and thus evolving in non-equilibrium, a Berry phase is acquired and a non-zero Berry flux and thus a non-zero Chern number are characterizing the topological band structure as to be *non-trivial*. For one-dimensional, models [61,71] with the variation of the external excitation frequency Ω_L replica of Floquet bands with a quantized change of the Berry phase $\gamma = \pi$ emerge in the spectrum. In three dimensions, the Berry phase is associated with the Wyckoff positions of the crystal and the Brillouin zone [61], and, as such, it cannot be derived by the pure form of the DMFT. $\vec{k}$-dependent information can be derived by so-called real-space or cluster DMFT solutions (R-DMFT or CDMFT) [72–75]; however, they have not been generalized to the non-equilibrium for three-dimensional systems. It is important to note that the system out of equilibrium acquires a non-zero Berry phase and Coulomb interactions lead to a Mott-type gap that closes due to the superposition by crossing Floquet bands; however, the opening of non-equilibrium induced Mott-gap replica can also be found for $\Omega_L = 0.95$ eV. The replica are complete at $\Omega_L = 1.9$ eV; see Figure 6a. For the increase of $\vec{E}_0$, these gap replica are again intersected by the next order of Floquet sidebands. The closing of the Mott-gap and the opening of side Mott-gaps, in the spectrum due to topological excitation, are classified as *non-trivial* topological effects. In Figure 7a, we present results of the LDOS for the same system of optically excited cubic ZnO rocksalt excited by an external laser energy of 1.75 eV and an increasing excitation intensity, Figure 7b shows the corresponding inverse lifetime $\Im\Sigma^R$ and Figure 7c shows the corresponding non-equilibrium distribution of electrons F^{neq}. In addition, for very small excitation intensities, the result for the non-equilibrium distribution function F^{neq} shows a profound deviation from the Fermi step in equilibrium. These occupied non-equilibrium states have a finite lifetime, especially at the inner band edges, which is a sign of the Franz-Keldysh effect [76–79], here in the sense of a topological effect, which is accessible in a pump-probe experiment.

The change of the polarization of the external excitation modifies the physical situation and the result. In particular, circular and elliptically polarized light can be formally written as a superposition of linear polarized waves. Thus, in the pure uncorrelated case, $U = 0$, one could think that the setup can be formally implemented in the sense of coupled matrices. In the strongly correlated system at hand, the physics is fundamentally different. The solution for the strongly correlated case, $U \neq 0$, in the non-equilibrium, including DMFT, will become more sophisticated since the coupled matrices will result in the entanglement of processes in some sense. This can be deduced from the result in Figure 7b, which displays the modification of non-equilibrium life-times of electronic states due to the varying excitation amplitudes. Such a modification is also qualitatively found for varying excitation frequencies.

The classification of correlated topological systems is an active research field [26,80,81]. At this point, we refer to Section 3.3 in this article, where we show, in our theoretical results the analysis of the single Floquet modes. For the investigation of the LDOS and the occupation number F^{neq}, as well as for the lifetimes of the non-equilibrium states, an artificial cut-off of the Floquet series, as it is described in the literature, does not make sense from the numerical physics point of view of DMFT in frequency space. This would hurt basically conservation laws and the cut-off would lead to a drift of the overall energy of the system; see Section 3.3. However, according to the bulk-boundary correspondence [2,3], the results of this work for bulk will be observed in a pump-probe experiment at the surface of the semiconductor sample. In the following, we discuss the development of Floquet topological states for an increasing external driving frequency Ω_L, see Figure 6, and, for an increasing amplitude of the driving, see Figure 7.

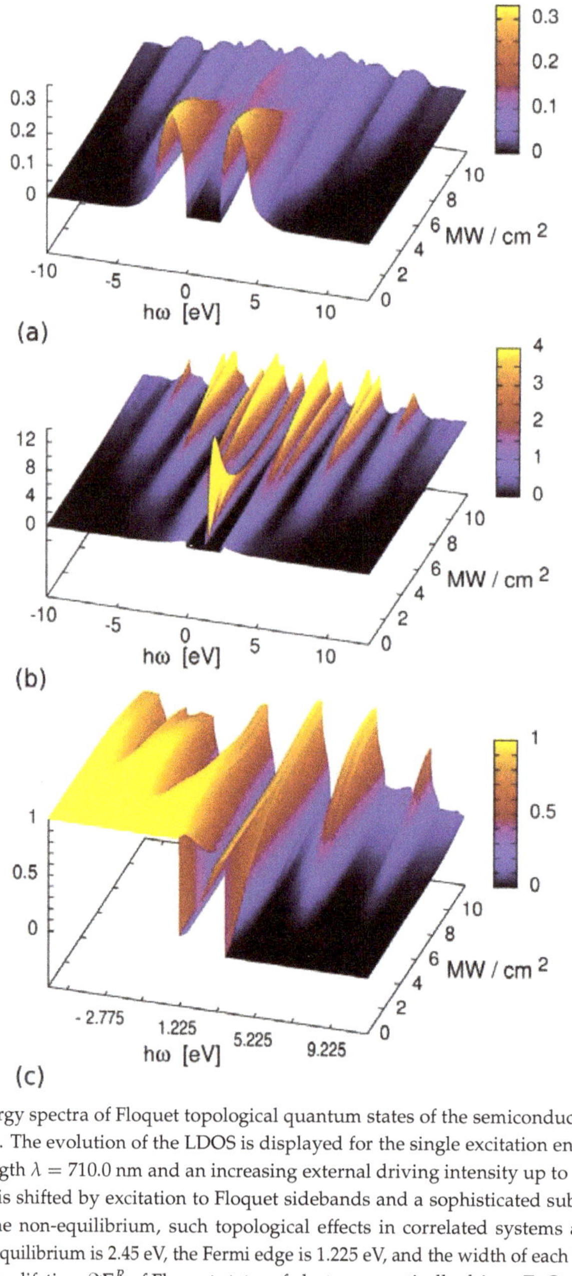

Figure 7. (**a**) energy spectra of Floquet topological quantum states of the semiconductor bulk in the non-equilibrium. The evolution of the LDOS is displayed for the single excitation energy of $\hbar\Omega_L =$ 1.75 eV, wavelength $\lambda = 710.0$ nm and an increasing external driving intensity up to 10.0 MW/cm^2. Spectral weight is shifted by excitation to Floquet sidebands and a sophisticated sub gab structure is formed. In the non-equilibrium, such topological effects in correlated systems are non-trivial. The bandgap in equilibrium is 2.45 eV, the Fermi edge is 1.225 eV, and the width of each band is 2.45 eV as well; (**b**) inverse lifetime $\Im\Sigma^R$ of Floquet states of electromagnetically driven ZnO rocksalt bulk in the non-equilibrium; (**c**) non-equilibrium distribution function F^{neq} of electrons in optically driven bulk ZnO rocksalt. Parameters in (**b**,**c**) are identical to (**a**). For a detailed discussion, please see Section 3.

3.2. Topological Generation of Higher Harmonics and of Optical Transparency

When we increase the external excitation energy of the system Ω_L from 0 eV to 4.0 eV, Floquet topological quantum states as well as the topologically induced Floquet band gaps for bulk matter are developed. Both valence and conduction band split in a multitude of Floquet sub-bands which cross each other. In Figure 6a, the evolution of a very clear Floquet fan for the valence as well as for the conduction band of the correlated matter in the non-equilibrium is found. When the excitation energy is increased up to 0.45 eV, the original band gap is subsequently closing, and the first crossing point in the semiconductor gap along the Fermi edge is found at 0.45 eV. With the increase of the excitation intensity, see Figure 6b, higher order Floquet sidebands are gaining spectral weight, and we find the next prominent crossing point at the Fermi edge for 0.2 eV. Band edges of higher order Floquet bands form crossing points with those of the first order. For an excitation energy of 0.42 eV, the crossing points of the first side bands (02) with the higher number side bands are found at the atomic energy of 1.08 eV and 2.3 eV, so above the valence band edge and deep in the gap of the semiconductor. Semiconductors are well known for fundamental absorption at the band edge of the valence band. We find here that the absorption coefficient of the semiconductor is topologically modulated. Non-trivial transitions at the crossing points of Floquet-valence subbands and Floquet-conduction subbands become significant. A higher order Floquet subband is usually physically reached by absorption or generation of higher harmonic procedures and we find a high probability for a topologically induced direct transitions from the fundamental to higher order bands for those points in the spectrum where a Floquet band edge intersects with the inner band edge of the equilibrium valence band. At any band edge directional scattering can be expected if the lifetimes of states are of a value that is applicable to the expected scattering processes. In general, the optical refractive index is topologically modulated, and electromagnetically induced transparency will become observable for intense excitations. The topologically induced Floquet bands overlap and cross each other. Consequentially, very pronounced features and narrow subgaps are formed in the LDOS, which correspond with sharp spikes in the expected life-times in the non-equilibrium. Floquet replica of valence and conduction bands are formed and the dispersion is renormalized. We also find regions for excitation energies from $\hbar\Omega_L = 0.5$ eV up to $\hbar\Omega_L = 0.85$ eV and from $\hbar\Omega_L = 1.1$ eV up to $\hbar\Omega_L = 1.45$ eV, which can be interpreted as a topologically induced metallic phase. These states are the result of the Franz-Keldysh effect [76–79] or AC-Stark effect, which is well known for high intensity excitation of semiconductor bulk and quantum wells [82,83].

From the viewpoint of correlated electronics in the non-equilibrium, we interpret our results as follows. For finite excitation frequencies, an instantaneous transition to the topologically induced Floquet band structure and a renormalized dispersion is derived. In the bulk system clear Floquet bands develop, if the sample is excited by an intense electrical field. This is observable in Figure 6.

In Figure 7, we display the same system as in Figure 6 for constant driving energy of 1.75 eV and an increasing driving intensity up to 10.0 MW/cm^2. We find the development of side bands and an overall vanishing semiconductor gap is found, which marks the transition from the semi-conductor to the topologically highly variable and switchable conductor in the non-equilibrium.

In this article, we do not investigate the coupling to a geometrical edge or a resonator mode. This will lead in the optical case to additional contributions in Equation (1) for the mode itself $\hbar\omega_0 a^\dagger a$ and the coupling term of the resonator or edge mode to the electron system of the bulk $g\sum_{i,\sigma} c^\dagger_{i,\sigma} c_{i,\sigma}(a^\dagger + a)$. $a^\dagger$ and a are the creator and the annihilator of the photon, and g is the variable coupling strength of the photonic mode to the electronic system [84]. From our results, here we can conclude already that, for semi-conductor cavities and quantum wells as well as for structures which enhance so called edge states, these geometrical edge or surface resonances will induce an additional topological effect within the full so far excitonic spectrum. It is an additional effect that occurs beyond the bulk boundary correspondence. Dressed states may release energy quanta, e.g., light, or an electronic current into the resonator component [38]. Thus, we expect from our results that such modes may become a sensible switch in non-equilibrium.

It can be expected as well that novel topological effects in the non-equilibrium occur from the geometry. If the energy of the system is conserved, these modes will have always an influence on the full spectrum of the LDOS, when the system is otherwise periodic in space and time. Thus, it is to clarify whether such modes may be of technological use. For the investigation of ZnO as a laser material, the influences of surface resonators will be subject to further investigations. It is on target to find out all the signatures of a topologically protected edge mode in correlated and strongly correlated systems out of equilibrium, and to classify the significance of topological effects for the occurrence of the electro-optical Kerr effect, the magneto-optical Kerr effect (MOKE) or the surface magneto-optical Kerr effect (SMOKE). We believe that in correlated many-body systems out of equilibrium a bulk boundary correspondence is given and will be experimentally found. Those results become modified or enhanced by a coupling of bulk states with the geometry of a micro- or a nanostructure and their geometrical resonances.

3.3. Consistency of the Numerical Framework

The consistency of the numerical formalism is generally checked by the sum over all Floquet indices with the physical meaning that energy conservation must be guaranteed in the non-equilibrium. Consequentially, we do not take into account thermalization procedures and the system's temperature remains constant. The analysis of the numerical validity as the normalized and frequency integrated density of states

$$N_i(\Omega_L) := \int \mathrm{d}\omega N(\omega, \Omega_L) = 1 \tag{9}$$

is confirmed in this work for summing over Floquet indices up to the order of 10. We discuss in Figure 8 on the l.h.s. the Floquet contributions with increasing number in steps of $n = 0, 2, 4, 6, 8, 10$. With an increasing order of the Floquet index, the amplitude of the Floquet contribution decreases towards the level of numerical precision of the DMFT self-consistency. This is definitely reached for $n = 10$ and thus it is the physical argument to cut the Floquet expansion off for $n = 10$. As a systems requirement, the Floquet contributions $G_{0\pm n}$ are perfectly mirror symmetric with respect to the Fermi edge, whereas the sum of both contributions is directly symmetric with respect to the Fermi edge. These symmetries are generally a proof of the validity of the numerical Fourier transformation and the numerical scheme. The order of magnitude of each Floquet contribution with a higher order than $n = 4$ is almost falling consistently with the rising Floquet index. We display results for the external laser wavelength of $\lambda = 710.0$ nm and the laser intensity of 3.8 MW/cm^2; the ZnO gap is assumed to be 2.45 eV, which is ZnO rocksalt as a laser active material. We include Floquet contributions up to a precision of 10^{-3} with regard to their effective difference from the final result on the r.h.s of Figure 8 as the sum to the nth-order. It corresponds to the accuracy of the self-consistent numerics.

The Floquet contributions, Figure 8, as such consequentially do not have a direct physical interpretation, however, the sum of all contributions is the local density of states, the LDOS, as a material characteristics. Whereas the lowest order Floquet contribution, compare Equation (2), G_{00} is symmetric to the Fermi edge but strictly positive, higher order contributions $G_{0\pm n}$ are mirror symmetric to each other and in sum they can have negative contributions to the result of the LDOS. The order of the Floquet contribution n numbers the evolving Floquet side bands which emerge in the LDOS, compare Figure 6a. The increase of mathematical and numerical precision has direct consequences for the finding and the accuracy of physical results, and the investigation of the coupling of the driven electronic system of the bulk with edge and surface modes will therefore profit. Bulk-surface coupling effects in nanostructure and waveguides are of great technological importance and the advantage of this numerical approach in contrast to time dependent DMFT frameworks in this respect is obvious.

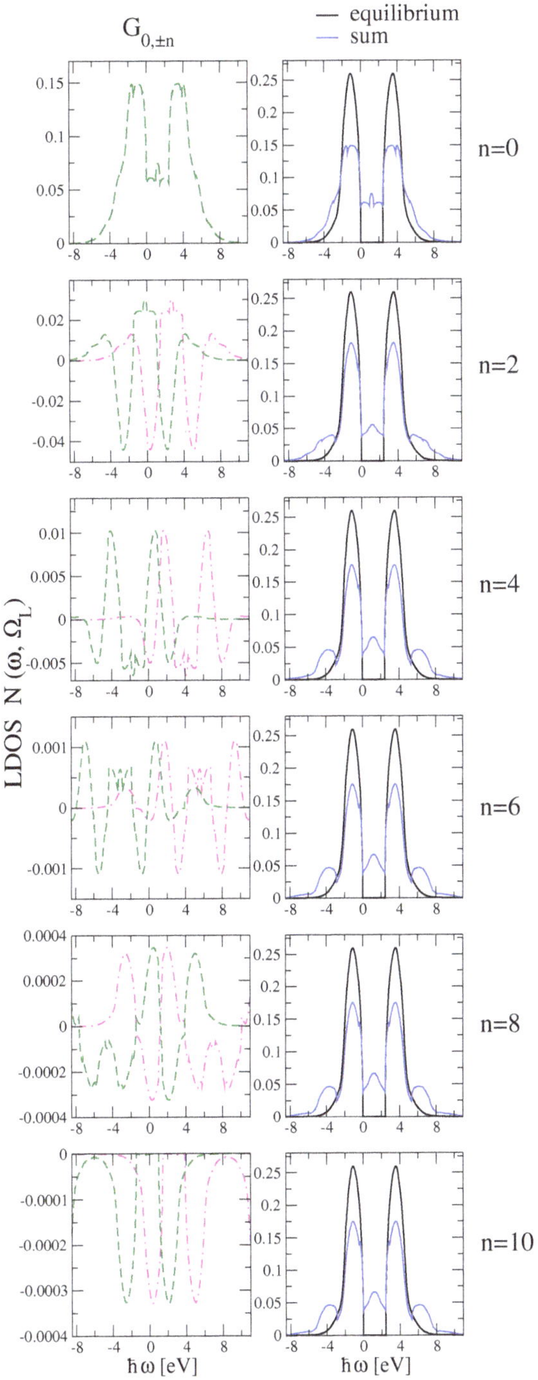

Figure 8. Floquet contributions and accuracy check of the numerical results for the LDOS of driven semiconductor bulk. The bandgap in equilibrium is 2.45 eV, the Fermi edge is 1.225 eV. For discussion, please see Section 3.3.

4. Conclusions

We investigated in this article the development of Floquet topological quantum states in wide band gap semiconductor bulk as a correlated electronic system with a generalized Hubbard model and with dynamical mean field theory in the non-equilibrium. We found that optical excitations induce a non-trivial band structure and, in several frequency ranges, a topologically induced metal phase is found as a result of the AC-Stark effect. The intersection of Floquet bands and band edges induces novel transitions, which may lead to up- and downconversion effects as well as to higher harmonic generation. The semiconductor absorption coefficient is topologically modulated. Non-trivial transitions at the crossing points of the underlying equilibrium band structure with the intersecting Floquet fans become possible and their efficiency is depending on the excitation power. We also find the development of pronounced novel sub gaps as areas of electromagnetically induced transparency. We also presented a consistency check as a physical consequence of the Floquet sum, which ensures numerically energy conservation. Our results for semiconductor bulk can be tested optoelectronic and magneto-optoelectronic experiments; they may serve as a guide towards innovative laser systems. The bulk semiconductor under topological non-equilibrium excitations as such has to be reclassified. It will be of great interest to investigate the interplay of topological bulk effects with additional surface resonances, a polariton coupling, or a surface magneto-optical modulation.

Author Contributions: Both authors equally contributed to the presented work. Both authors were equally involved in the preparation of the manuscript. Both authors have read and approved the final manuscript.

Funding: This research received no external funding.

Acknowledgments: The authors thank H. Monien, H. Wittel and F. Hasselbach for highly valuable discussions.

Conflicts of Interest: The authors declare that there are no conflict of interests.

References

1. Kosterlitz, J.M.; Thouless, D.J. Ordering, metastability and phase transitions in two-dimensional systems. *J. Phys. C Solid State Phys.* **1973**, *6*, 1181. [CrossRef]
2. Fu, L.; Kane, C.L.; Mele, E.J. Topological Insulators in Three Dimensions. *Phys. Rev. Lett.* **2007**, *98*, 106803. [CrossRef]
3. Hasan, M.Z.; Kane, C.L. Colloquium: Topological insulators. *Rev. Mod. Phys.* **2010**, *82*, 3045. [CrossRef]
4. Fu, L.; Kane, C.L. Superconducting Proximity Effect and Majorana Fermions at the Surface of a Topological Insulator. *Phys. Rev. Lett.* **2008**, *100*, 096407. [CrossRef] [PubMed]
5. Moore, G.; Read, N. Nonabelions in the fractional quantum hall effect. *Nucl. Phys. B* **1991**, *360*, 362. [CrossRef]
6. Goldmann, N.; Dalibard, J. Periodically Driven Quantum Systems: Effective Hamiltonians and Engineered Gauge Fields. *Phys. Rev. X* **2014**, *4*, 031027. [CrossRef]
7. Zutic, I.; Fabian, J.; Das Sarma, S. Spintronics: Fundamentals and applications. *Rev. Mod. Phys.* **2004**, *76*, 323. [CrossRef]
8. Nayak, C.; Simon, S.H.; Stern, A.; Freedman, M.; Das Sarma, S. Non-Abelian anyons and topological quantum computation. *Rev. Mod. Phys.* **2008**, *80*, 1083. [CrossRef]
9. Rechtsman, M.C.; Zeuner, J.M.; Plotnik, Y.; Lumer, Y.; Podolsky, D.; Dreisow, F.; Nolte, S.; Segev, M.; Szameit, A. Photonic Floquet topological insulators. *Nature* **2013**, *496*, 196. [CrossRef]
10. Bandres, M.A.; Wittek, S.; Harari, G.; Parto, M.; Ren, J.; Segev, M.; Christodoulides, D.N.; Khajavikhan, M. Topological insulator laser: Experiments. *Science* **2018**, *359*. [CrossRef]
11. Lubatsch, A.; Frank, R. A Self-Consistent Quantum Field Theory for Random Lasing. *Appl. Sci.* **2019**, *9*, 2477. [CrossRef]
12. Bernevig, B.A.; Hughes, T.L.; Zhang, S.-C. Quantum Spin Hall Effect and Topological Phase Transition in HgTe Quantum Wells. *Science* **2006**, *314*, 1757. [CrossRef] [PubMed]
13. König, M.; Wiedmann, S.; Brüne, C.; Roth, A.; Buhmann, H.; Molenkamp, L.W.; Qi, X.-L.; Zhang, S.-C. Quantum spin hall insulator state in HgTe quantum wells. *Science* **2007**, *318*, 766. [CrossRef] [PubMed]

14. Hsieh, D.; Qian, D.; Wray, L.; Xia, Y.; Hor, Y.S.; Cava, R.J.; Hasan, M.Z. A topological Dirac insulator in a quantum spin Hall phase: Experimental observation of first strong topological insulator. *Nature* **2008**, *452*, 970–974. [CrossRef] [PubMed]
15. Xia, Y.; Qian, D.; Hsieh, D.; Wray, L.; Pal, A.; Lin, H.; Bansil, A.; Grauer, D.; Hor, Y.S.; Cava, R.J.; et al. Discovery (theoretical prediction and experimental observation) of a large-gap topological-insulator class with spin-polarized single-Dirac-cone on the surface. *Nat. Phys.* **2009**, *5*, 398. [CrossRef]
16. Zhang, H.; Liu, C.-X.; Qi, X.-L.; Dai, X.; Fang, Z.; Zhang, S.-C. Topological insulators in Bi_2Se_3, Bi_2Te_3 and Sb_2Te_3 with a single Dirac cone on the surface. *Nat. Phys.* **2009**, *5*, 438–442. [CrossRef]
17. Lindner, N.H.; Refael, G.; Galitski, V. Floquet Topological Insulator in Semiconductor Quantum Wells. *Nat. Phys.* **2011**, *7*, 490–495. [CrossRef]
18. Katan, Y.T.; Podolsky, D. Modulated Floquet Topological Insulators. *Phys. Rev. Lett.* **2013**, *110*, 016802. [CrossRef]
19. Rudner, M.S.; Lindner, N.H.; Berg, E.; Levin, M. Anomalous Edge States and the Bulk-Edge Correspondence for Periodically Driven Two-Dimensional Systems. *Phys. Rev. X* **2013**, *3*, 031005. [CrossRef]
20. Wang, Y.; Liu, Y.; Wang, B. Effects of light on quantum phases and topological properties of two-dimensional Metal-organic frameworks. *Sci. Rep.* **2016**, *4*, 41644. [CrossRef]
21. Kitagawa, T.; Oka, T.; Brataas, A.; Fu, L.; Demler, E. Transport properties of nonequilibrium systems under the application of light: Photoinduced quantum Hall insulators without Landau levels. *Phys. Rev. B* **2011**, *84*, 235108. [CrossRef]
22. Gu, Z.; Fertig, H.A.; Arovas, D.P.; Auerbach, A. Floquet Spectrum and Transport through an Irradiated Graphene Ribbon. *Phys. Rev. Lett.* **2011**, *107*, 216601. [CrossRef] [PubMed]
23. Wang, Y.H.; Steinberg, H.; Jarillo-Herrero, P.; Gedik, N. Observation of Floquet-Bloch States on the Surface of a Topological Insulator. *Science* **2013**, *342*, 453–457. [CrossRef] [PubMed]
24. Lindner, N.H.; Bergman, D.L.; Refael, G.; Galitski, V. Topological Floquet spectrum in three dimensions via a two-photon resonance. *Phys. Rev. B* **2013**, *87*, 235131. [CrossRef]
25. Jiang, L.; Kitagawa, T.; Alicea, J.; Akhmerov, A.R.; Pekker, D.; Refael, G.; Cirac, J.I.; Demler, E.; Lukin, M.D.; Zoller, P. Majorana Fermions in Equilibrium and in Driven Cold-Atom Quantum Wires. *Phys. Rev. Lett.* **2011**, *106*, 220402. [CrossRef] [PubMed]
26. D'Alessio, L. Rigol, Dynamical preparation of Floquet Chern insulators. *Nat. Commun.* **2015**, *6*, 8336. [CrossRef]
27. Grushin, A.G.; Gomez-Leon, A.; Neupert, T. Floquet Fractional Chern Insulators. *Phys. Rev. Lett.* **2014**, *112*, 156801. [CrossRef]
28. Bergholtz, E.; Liu, Z. Topological Flat Band Models and Fractional Chern Insulators. *Int. J. Mod. Phys. B* **2013**, *27*, 1330017. [CrossRef]
29. Floquet, G. Sur les équations différentielles linéaires à coefficients périodiques. *Ann. l' Ecole Norm. Sup.* **1883**, *12*, 47–88. [CrossRef]
30. Razavi-Khosroshahi, H.; Edalati, K.; Wu, J.; Nakashima, Y.; Arita, M.; Ikoma, Y.; Sadakiyo, M.; Inagaki, Y.; Staykov, A.; Yamauchi, M.; et al. High-pressure zinc oxide phase as visible-light-active photocatalyst with narrow band gap. *J. Mater. Chem. A* **2017**, *5*, 20298–20303. [CrossRef]
31. Fritsch, D.; Schmidt, H.; Grundmann, M. Pseudopotential band structures of rocksalt MgO, ZnO, and Mg 1 x Zn x O. *Appl. Phys. Lett.* **2006**, *88*, 134104. [CrossRef]
32. Dixit, H.; Saniz, R.; Lamoen, D.; Partoens, B. The quasiparticle band structure of zincblende and rocksalt ZnO. *J. Phys. Condens. Matter* **2010**, *22*, 125505. [CrossRef] [PubMed]
33. Huang, F.; Lin, Z.; Lin, W.; Zhang, J.; Ding, K.; Wang, Y.; Zheng, Q.; Zhan, Z.; Yan, F.; Chen, D.; et al. Research progress in ZnO single-crystal: Growth, scientific understanding, and device applications. *Chin. Sci. Bull.* **2014**, *59*, 1235–1250. [CrossRef]
34. Park, W.I.; Jun, Y.H.; Jung, S.W.; Yi, G.-C. Exciton emissions observed in ZnO single crystal nanorods. *Appl. Phys. Lett.* **2003**, *82*, 964–966. [CrossRef]
35. Faisal, F.H.M.; Kaminski, J.Z. Floquet-Bloch theory of high-harmonic generation in periodic structures. *Phys. Rev. A* **1997**, *56*, 748. [CrossRef]
36. Lubatsch, A.; Frank, R. Evolution of Floquet topological quantum states in driven semiconductors. *Eur. Phys. J. B* **2019**, *92*, 215. [CrossRef]

37. Frank, R. Quantum criticality and population trapping of fermions by non-equilibrium lattice modulations. *New J. Phys.* **2013**, *15*, 123030. [CrossRef]
38. Frank, R. Coherent control of Floquet-mode dressed plasmon polaritons. *Phys. Rev. B* **2012**, *85*, 195463. [CrossRef]
39. Frank, R. Non-equilibrium polaritonics - Nonlinear effects and optical switching. *Ann. Phys.* **2013**, *525*, 66–73. [CrossRef]
40. Grifoni, M.; Hänggi, P. Driven quantum tunneling. *Phys. Rep.* **1998**, *304*, 229–354. [CrossRef]
41. Restrepo, S.; Cerrillo, J.; Bastidas, V.M.; Angelakis, D.G.; Brandes, T. Driven Open Quantum Systems and Floquet Stroboscopic Dynamics. *Phys. Rev. Lett.* **2016**, *117*, 250401. [CrossRef]
42. Eckardt, A. Colloquium: Atomic quantum gases in periodically driven optical lattices. *Rev. Mod. Phys.* **2017**, *89*, 011004-1. [CrossRef]
43. Kalthoff, M.H.; Uhrig, G.S.; Freericks, J.K. Emergence of Floquet behavior for lattice fermions driven by light pulses. *Phys. Rev. B* **2018**, *98*, 035138. [CrossRef]
44. Sentef, M.A.; Claassen, M.; Kemper, A.F.; Moritz, B.; Oka, T.; Freericks, J.K.; Devereaux, T.P. Theory of Floquet band formation and local pseudospin textures in pump-probe photoemission of graphene. *Nat. Commun.* **2015**, *6*, 7047. [CrossRef] [PubMed]
45. Yuan, L.; Fan, S. Topologically non-trivial Floquet band structure in a system undergoing photonic transitions in the ultra-strong coupling regime. *Phys. Rev. A* **2015**, *92*, 053822. [CrossRef]
46. Lubatsch, A.; Kroha, J. Optically driven Mott-Hubbard systems out of thermodynamical equilibrium. *Ann. Phys.* **2009**, *18*, 863–867. [CrossRef]
47. Georges, A.; Kotliar, G.; Krauth, W.; Rozenberg, M.J. Dynamical mean-field theory of strongly correlated fermion systems and the limit of infinite dimensions. *Rev. Mod. Phys.* **1996**, *68*, 13. [CrossRef]
48. Metzner, W.; Vollhardt, D. Correlated Lattice Fermions in $d = \infty$ Dimensions. *Phys. Rev. Lett.* **1989**, *62*, 324. [CrossRef] [PubMed]
49. Schmidt, P.; Monien, H. Nonequilibrium dynamical mean-field theory of a strongly correlated system. *arXiv* **2002**, arXiv:0202046.
50. Maier, T.; Jarrell, M.; Pruschke, T.; Hettler, M.H. Quantum cluster theories. *Rev. Mod. Phys.* **2005**, *77*, 1027. [CrossRef]
51. Freericks, J.K.; Turkowski, V.M.; Zlatic, V. Nonequilibrium Dynamical Mean-Field Theory. *Phys. Rev. Lett.* **2006**, *97*, 266408. [CrossRef] [PubMed]
52. Tsuji, N.; Oka, T.; Aoki, H. Nonequilibrium Steady State of Photoexcited Correlated Electrons in the Presence of Dissipation. *Phys. Rev. Lett.* **2009**, *103*, 047403. [CrossRef] [PubMed]
53. Lin, N.; Marianetti, C.A.; Millis, A.J.; Reichman, D.R. Dynamical Mean-Field Theory for Quantum Chemistry. *Phys. Rev. Lett.* **2011**, *106*, 096402. [CrossRef] [PubMed]
54. Zgid, D.; Chan, G.K.-L. Dynamical mean-field theory from a quantum chemical perspective. *J. Chem. Phys.* **2011**, *134*, 094115. [CrossRef] [PubMed]
55. Aoki, H.; Tsuji, N.; Eckstein, M.; Kollar, M.; Oka, T.; Werner, P. Nonequilibrium dynamical mean-field theory and its applications. *Rev. Mod. Phys.* **2014**, *86*, 779. [CrossRef]
56. Frank, R. Population trapping and inversion in ultracold Fermi gases by excitation of the optical lattice-Non-equilibrium Floquet-Keldysh description. *Appl. Phys. B* **2013**, *113*, 41–47. [CrossRef]
57. Sorantin, M.E.; Dorda, A.; Held, K.; Arrigoni, E. Impact ionization processes in the steady state of a driven Mott-insulating layer coupled to metallic leads. *Phys. Rev. B* **2018**, *97*, 115113. [CrossRef]
58. Hofstetter, W.; Qin, T. Topological singularities and the general classification of Floquet-Bloch systems. *J. Phys. B At. Mol. Opt. Phys.* **2018**, *51*, 082001. [CrossRef]
59. Qin, T.; Hofstetter, W. Nonequilibrium steady states and resonant tunneling in time-periodically driven systems with interactions. *Phys. Rev. B* **2018**, *97*, 125115. [CrossRef]
60. Bukov, M.; D'Alessio, L.; Polkovnikov, A. Universal High-Frequency Behavior of Periodically Driven Systems: From Dynamical Stabilization to Floquet Engineering. *Adv. Phys.* **2015**, *64*, 139–226. [CrossRef]
61. Zak, J. Berry's phase for energy bands in solids. *Phys. Rev. Lett.* **1989**, *62*, 2747–2750. [CrossRef]
62. Xiao, D.; Chang, M.-C.; Niu, Q. Berry phase effects on electronic properties. *Rev. Mod. Phys.* **2009**, *82*, 1959–2007. [CrossRef]

63. Gresch, D.; Autes, G.; Yazyev, O.V.; Troyer, M.; Vanderbilt, D.; Bernevig, B.A.; Soluyanov, A.A. Z2Pack: Numerical implementation of hybrid Wannier centers for identifying topological materials. *Phys. Rev. B* **2017**, *95*, 075146. [CrossRef]
64. Chang, P.-C.; Lu, J.G. Temperature dependent conduction and UV induced metal-to-insulator transition in ZnO nanowires. *Appl. Phys. Lett.* **2008**, *92*, 212113. [CrossRef]
65. Chang, P.-C.; Chien, C.-J.; Stichtenoth, D.; Ronning, C.; Lu, J.G. Finite size effect in ZnO nanowires. *Appl. Phys. Lett.* **2007**, *90*, 113101. [CrossRef]
66. Koster, R.S.; Changming, M.F.; Dijkstra, M.; van Blaaderen, A.; van Huis, M.A. Stabilization of Rock Salt ZnO Nanocrystals by Low-Energy Surfaces and Mg Additions: A First-Principles Study. *J. Phys. Chem. C* **2015**, *119*, 5648–5656. [CrossRef]
67. Wegener, M. *Extreme Nonliner Optics*; Springer: Berlin/Heidelberg, Germany, 2004; ISBN 3-540-22291-X.
68. King-Smith, R.D.; Vanderbilt, D. Theory of polarization of crystalline solids. *Phys. Rev. B* **1993**, *47*, 1651–1654. [CrossRef]
69. Vanderbilt, D.; King-Smith, R.D. Electric polarization as a bulk quantity and its relation to surface charge. *Phys. Rev. B* **1993**, *48*, 4442. [CrossRef]
70. Resta, R. Macroscopic polarization in crystalline dielectrics: The geometric phase approach. *Rev. Mod. Phys.* **1994**, *66*, 899–915. [CrossRef]
71. Dal Lago, V.; Atala, M.; Foa Torres, L.E.F. Floquet topological transitions in a driven one-dimensional topological insulator. *Phys. Rev. A* **2015**, *92*, 023624. [CrossRef]
72. Gull, E.; Millis, A.; Lichtenstein, A.I.; Rubtsov, A.N.; Troyer, M.; Werner, P. Continuous-time Monte Carlo methods for quantum impurity models. *Rev. Mod. Phys.* **2011**, *83*, 349. [CrossRef]
73. Snoek, M.; Titvinidze, I.; Toke, C.; Byczuk, K.; Hofstetter, W. Antiferromagnetic order of strongly interacting fermions in a trap: Real-space dynamical mean-field analysis. *New J. Phys.* **2008**, *10*, 093008. [CrossRef]
74. Peters, R.; Yoshida, T.; Sakakibara, H.; Kawakami, N. Coexistence of light and heavy surface states in a topological multiband Kondo insulator. *Phys. Rev. B* **2016**, *93*, 235159. [CrossRef]
75. Peters, R.; Yoshida, T.; Kawakami, N. Magnetic states in a three-dimensional topological Kondo insulator. *Phys. Rev. B* **2018**, *98*, 075104. [CrossRef]
76. Franz, W. Einfluß eines elektrischen Feldes auf eine optische Absorptionskante. *Z. Naturforschung A* **1958**, *13*, 484–489. [CrossRef]
77. Keldysh, L.V. The Effect of a Strong Electric Field on the Optical Properties of Insulating Crystals. *J. Exptl. Theoret. Phys.* **1957**, *33*, 994–1003; reprinted in *Soviet Phys. JETP* **1958**, *6*, 763–770. Available online: jetp.ac.ru/cgi-bin/dn/e_007_05_0788.pdf (accessed on 4 October 2019).
78. Keldysh, L.V. Ionization in the Field of a Strong Electromagnetic Wave. *J. Exptl. Theoret. Phys.* **1964**, *47*, 1945–1957; reprinted in *Soviet Phys. JETP* **1964**, *20*, 1307–1314. Available online: jetp.ac.ru/cgi-bin/dn/e_020_05_1307.pdf (accessed on 4 October 2019).
79. Ivanov, A.L.; Keldysh, L.V. ; Panashchenko, V.V. Low-threshold exciton-biexciton optical Stark effect in direct-gap semiconductors. *Zh. Eksp. Teor. Fiz.* **1991**, *99*, 641–658.
80. Manmana, S.; Essin, A.M.; Noack, R.M.; Gurarie, V. Topological invariants and interacting one-dimensional fermionic systems. *Phys. Rev. B* **2012**, *86*, 205119. [CrossRef]
81. Rachel, S. Interacting topological insulators: A review. *Rep. Prog. Phys.* **2018**, *81*, 116501. [CrossRef]
82. Miller, D.A.B.; Chemla, D.S.; Damen, T.C.; Gossard, A.C.; Wiegmann, W.; Wood, T.H.; Burrus, C.A. Band-Edge Electroabsorption in Quantum Well Structures: The Quantum-Confined Stark Effect. *Phys. Rev. Lett.* **1984**, *53*, 2173–2176. [CrossRef]
83. Chemla, D.S.; Knox, W.H.; Miller, D.A.B.; Schmitt-Rink, S.; Stark, J.B.; Zimmermann, R. The excitonic optical stark effect in semiconductor quantum wells probed with femtosecond optical pulses. *J. Lumin.* **1989**, *44*, 233–246. [CrossRef]
84. Forn-Diaz, P.; Lamata, L.; Rico, E.; Kono, J.; Solano, E. Ultrastrong coupling regimes of light-matter interaction. *Rev. Mod. Phys.* **2019**, *91*, 025005-1. [CrossRef]

symmetry

MDPI

Article

Spin-Boson Model as A Simulator of Non-Markovian Multiphoton Jaynes-Cummings Models

Ricardo Puebla [1,*], Giorgio Zicari [1], Iñigo Arrazola [2], Enrique Solano [2,3,4], Mauro Paternostro [1] and Jorge Casanova [2,3]

1. Centre for Theoretical Atomic, Molecular, and Optical Physics, School of Mathematics and Physics, Queen's University Belfast, Belfast BT7 1NN, UK; gzicari01@qub.ac.uk (G.Z.); m.paternostro@qub.ac.uk (M.P.)
2. Department of Physical Chemistry, University of the Basque Country UPV/EHU, Apartado 644, E-48080 Bilbao, Spain; iarrazola003@gmail.com (I.A.); enr.solano@gmail.com (E.S.); jcasanovamar@gmail.com (J.C.)
3. IKERBASQUE, Basque Foundation for Science, María Díaz de Haro 3, 48013 Bilbao, Spain
4. Department of Physics, Shanghai University, Shanghai 200444, China

* Correspondence: r.puebla@qub.ac.uk

Received: 17 April 2019; Accepted: 14 May 2019; Published: 20 May 2019

Abstract: The paradigmatic spin-boson model considers a spin degree of freedom interacting with an environment typically constituted by a continuum of bosonic modes. This ubiquitous model is of relevance in a number of physical systems where, in general, one has neither control over the bosonic modes, nor the ability to tune distinct interaction mechanisms. Despite this apparent lack of control, we present a suitable transformation that approximately maps the spin-boson dynamics into that of a *tunable* multiphoton Jaynes-Cummings model undergoing dissipation. Interestingly, the latter model describes the coherent interaction between a spin and a single bosonic mode via the simultaneous exchange of n bosons per spin excitation. Resorting to the so-called reaction coordinate method, we identify a relevant collective bosonic mode in the environment, which is then used to generate multiphoton interactions following the proposed theoretical framework. Moreover, we show that spin-boson models featuring structured environments can lead to non-Markovian multiphoton Jaynes-Cummings dynamics. We discuss the validity of the proposed method depending on the parameters and analyse its performance, which is supported by numerical simulations. In this manner, the spin-boson model serves as a good analogue quantum simulator for the inspection and realization of multiphoton Jaynes-Cummings models, as well as the interplay of non-Markovian effects and, thus, as a simulator of light-matter systems with tunable interaction mechanisms.

Keywords: spin-boson model; Jaynes-Cummings model; multiphoton processes; quantum simulation

1. Introduction

The rapid technological progress we have experienced during the last few decades has made possible previously inconceivable experiments at the quantum regime, boosting their degree of precision, isolation and control to unprecedented limits [1]. Currently, quantum systems can be inspected in a very controllable manner in a number of distinct setups. This experimental breakthrough has therefore stimulated the emergence of research areas such as quantum information and computation and quantum simulation, where the exploitation of quantum effects will allow us to surpass both the capabilities of their classical counterparts in the near future [2]. In particular, quantum simulation considers a scenario in which a well-controlled quantum system serves as a simulator of other inaccessible systems [3–5]. In this manner, interesting quantum dynamics (i.e., the target dynamics) may be explored using, for example, optical lattices [6] or trapped ions [7]. The target dynamics can be obtained either by decomposing the time-evolution propagator in a set of simple quantum operations (digital quantum simulation) or by finding a map that brings the Hamiltonian into the desired form of the model to be simulated (analogous to quantum simulation) [5]. In this article, we will consider the latter method, by using as a quantum simulator the paradigmatic spin-boson model [8,9].

The spin-boson model describes a spin immersed in an environment formed by a large, typically infinite, number of bosonic modes, in contrast to the quantum Rabi or Jaynes-Cummings models where the interaction comprises a single bosonic mode [10–13]. The spin-boson model encompasses very rich physics depending on how the spin couples with the distinct bosonic modes. Hence, while it is a minimal model to scrutinize the quantum effects of dissipation, it has application in a broad range of systems [8,9], ranging from defects in solid state platforms to quantum emitters in biological systems [14]. Moreover, much effort inspecting the spin-boson model has dealt with its critical behaviour, that is with the emergence of a quantum phase transition between a delocalized and a localized phase of the spin degree of freedom as one increases the spin-environment coupling [8]. The simulation of the spin-boson model (or of a generic open quantum system) in the strong coupling regime is however computationally very demanding, as acknowledged in [15–21], since the spin and the bosonic modes become entangled, forming a truly quantum many-body system. In some situations, one can still resort to analytical methods, which may simplify the problem considerably. Among these methods one finds the so-called reaction coordinate mapping [22–28], which can be viewed as a first step of the more general semi-infinite chain mapping of the environmental degrees of freedom [29,30]. The reaction coordinate is defined as a collective mode of the original environment oscillators. In this manner, one can bring the spin-boson model into the form of a generalized quantum Rabi model [10,11,13] whose bosonic mode undergoes dissipation as it interacts with the residual environment. In particular cases, upon rearranging the original environmental degrees of freedom, the dissipation acquires a Markovian character, hence simplifying considerably the complexity of the problem (see for example [24]). It is also worth mentioning other attempts to capture quantum dynamics effectively with complex system-environment interactions, as for example the recent work relying on pseudo-modes [31], which builds on the proven equivalence for the dynamics of the system in both frames [32].

The quantum Rabi model (QRM), as well as its simplified version known as the Jaynes-Cummings model (JCM) [12] play a central role in the description of light-matter interacting systems and in quantum information science [2,13]. In these models, the interaction mechanism between the spin and bosonic degrees of freedom has a linear form, namely the spin gets excited or deexcited by absorbing or emitting one bosonic excitation. While this interaction is ubiquitous in quantum physics and with application in various experimental platforms [33], other forms of a spin-boson exchange mechanisms beyond this simple case are also of interest. On the one hand, interactions beyond the linear fashion are of relevance for several applications in quantum computation and simulation (e.g., the Kerr effect [34]). Furthermore, these exchange mechanisms may unveil interesting phenomena in light-matter systems [35,36], as well as in their multiple spin counterparts [37]. One possible generalization of the QRM or JCM consists of considering a spin-multiphoton interaction, where the spin exchanges n excitations simultaneously with the bosonic mode. Such a generalization is often regarded as n-photon QRM or JCM, (nQRM or nJCM), and it has recently attracted attention mainly in its $n = 2$ form [35,36,38–41], although models with $n > 2$ have been also analysed [42]. From an experimental point of view, however, such multiphoton terms are typically hard to attain. Thus, its realization may benefit from quantum simulation protocols, allowing for enough tunability and control over multiphoton interaction terms, as proposed using optical trapped ions [35,38] or superconducting qubits [40]. These latter schemes realize effective multiphoton exchange terms by exploiting the nonlinear fashion in which the spin and bosonic degrees of freedom couple. It is however still possible to realize such multiphoton models even when the setup comprises solely a linear, i.e., standard, interaction mechanism, and thus, it is not suited for a direct simulation of these models, as shown in [43].

In this article, we follow the theoretical framework developed in [43,44], combining the ideas of the reaction-coordinate mapping [22–28] to show that the paradigmatic spin-boson model, featuring a continuum of bosonic modes, can serve as an analogue quantum simulator for the realization of different dissipative multiphoton Jaynes-Cummings models by tuning the frequency and bias parameter of the spin. In this manner, we demonstrate the emergence of a connection between the dynamics of these paradigmatic and fundamental quantum models, which was not previously unveiled. Moreover, as the spin-boson model is of considerable experimental significance, i.e., it describes the ubiquitous scenario of a two-level system interacting with an arbitrary environment, our method paves the way for the simulation of multiphoton Jaynes-Cummings models in distinct setups. In particular, by considering a full spin-boson model, we naturally extend the theoretical framework beyond the standard local master equation description of dissipation effects in the simulator, as considered in [44]. Furthermore, we show that the simulated multiphoton Jaynes-Cummings models may acquire non-Markovian behaviours when the spin-boson model features a structured environment, thus highlighting the suitability of the proposed theoretical framework to explore aspects of non-Markovianity in distinct light-matter interacting systems.

The article is organized as follows. In Section 2, we introduce the spin-boson model, while in Section 3, we explain how to map the spin-boson model into a different Hamiltonian comprising the desired spin-multiphoton interaction terms and discuss how the dissipative effects must be transformed into the aimed model. For that, we first introduce the reaction coordinate mapping in Section 3.1, while in Section 3.2, we explain how to extend the theoretical framework to incorporate further bosonic modes in the realization of the desired multiphoton model. After having provided the theoretical derivation of how to perform the analogue quantum simulation, we present examples and numerical results for the simulation of different multiphoton Jaynes-Cummings models in Section 4. Finally, we summarize the main conclusions of this article in Section 5.

2. The Spin-Boson Model

The spin-boson model describes a two-level system interacting with a large, typically infinite, number of bosonic modes, which constitute the environment. This model has been acknowledged as a paradigm for the inspection of quantum dissipation and quantum-to-classical transition [8,9]. As many physical systems can be well approximated as a two-level system for sufficient low temperature, the spin-boson model has become a cornerstone in the description of quantum effects in diverse physical realizations, ranging from quantum-based setups [8,9] to biological complexes [14]. In addition, this model has played a key role in the development of the theory of open quantum systems [45], providing a suitable test-bed to benchmark distinct approximations and tools aimed to deal with the large number of environment degrees of freedom efficiently. Moreover, the relevance of the spin-boson model also encompasses the context of critical systems, as it features a quantum phase transition between spin localized and delocalized phases (see References [46,47] and the references therein). Hence, the spin-boson model exhibits rich physics, and it is of fundamental relevance in many different areas of research.

The Hamiltonian of the spin-boson model can be written as:

$$H_{\mathrm{SB}} = H_{\mathrm{S}} + H_{\mathrm{E}} + H_{\mathrm{S-E}} \tag{1}$$

where each contribution reads as:

$$H_{\rm S} = \frac{\epsilon_0}{2}\sigma_z + \frac{\Delta_0}{2}\sigma_x, \tag{2}$$

$$H_{\rm E} = \sum_k \omega_k c_k^\dagger c_k, \tag{3}$$

$$H_{\rm S-E} = \sigma_x \sum_k f_k (c_k + c_k^\dagger). \tag{4}$$

The first two terms represent the free-energy Hamiltonians of the spin and environment, while the last describes the interaction between them. Here, we consider that the frequency splitting of the spin is given by Δ_0, while ϵ_0 accounts for the bias between the eigenstates of the two-level system $|\pm\rangle$ and with $\vec{\sigma} = (\sigma_x, \sigma_y, \sigma_z)$ the usual spin-$\frac{1}{2}$ Pauli matrices (see Figure 1a). Hence, $\sigma_x |\pm\rangle = \pm |\pm\rangle$, $\sigma_z |e\rangle = |e\rangle$ and $\sigma_z |g\rangle = -|g\rangle$. The interaction with the environment is dictated by $H_{\rm S-E}$, where the $k^{\rm th}$ mode with energy ω_k is coupled to the spin with a strength f_k. These bosonic modes fulfil the usual commutation relation $[c_k, c_{k'}^\dagger] = \delta_{k,k'}$. Remarkably, the system-environment interaction can be completely characterized in terms of the spectral density, $J_{\rm SB}(\omega) = \sum_k f_k^2 \delta(\omega - \omega_k)$, which here is assumed to be known. In anticipation of the developed theoretical framework that allows us to bring $H_{\rm SB}$ into the form of a multiphoton Jaynes-Cummings model, we comment that while the frequency splitting Δ_0 tunes the multiphoton order of the interaction, the bias parameter ϵ_0 will be proportional to the interacting strength of the simulated model (see Section 3).

In addition, we comment that one could consider the application of n_d drivings onto the spin. As discussed in [43,44], under certain conditions that we will explain in the following section, applying spin drivings enables the simultaneous realization of different multiphoton Jaynes-Cummings interaction terms. In this manner, while a multiphoton Jaynes-Cummings model can be attained without the need for any driving, $n_d = 0$, the realization of a multiphoton quantum Rabi model requires the application of at least one, i.e., $n_d = 1$. In general, the free-energy Hamiltonian of the spin under n_d drivings with amplitude ϵ_j and detuning Δ_j with respect to the spin frequency splitting Δ_0 reads as:

$$H_{\rm S,d} = \frac{\Delta_0}{2}\sigma_x + \sum_{j=0}^{n_d} \frac{\epsilon_j}{2}\left[\cos(\Delta_j - \Delta_0)t\, \sigma_z + \sin(\Delta_j - \Delta_0)t\, \sigma_y\right]. \tag{5}$$

Clearly, setting $\epsilon_{j>0} = 0$ (or $\Delta_j = \Delta_0$), we recover the form of the standard drivingless $H_{\rm S}$ given in Equation (2). For the sake of simplicity, in this article, we will focus on cases with $n_d = 0$, i.e., aiming to realize multiphoton Jaynes-Cummings models. However, we stress that the procedure explained in the following can be applied in a straightforward manner when $n_d > 0$.

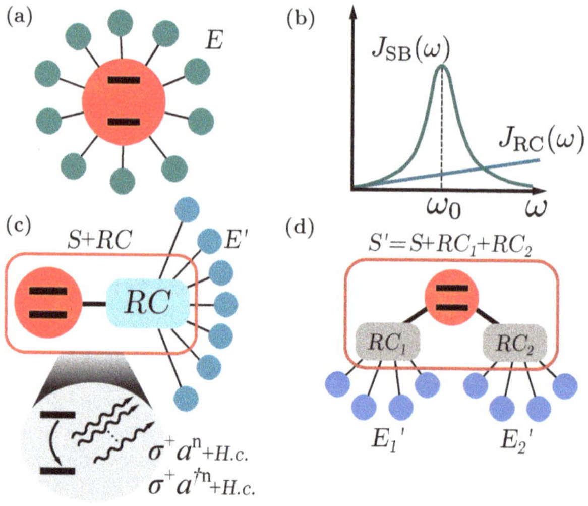

Figure 1. (**a**) Spin-boson model in the customary star configuration, where each of the circles corresponds to a harmonic oscillator of the environment with frequency ω_k interacting with the spin through $\sigma_x f_k(c_k + c_k^\dagger)$, before the reaction coordinate mapping. In (**b**), we show an underdamped spin-boson spectral density $J_{SB}(\omega)$, peaked at ω_0 (cf. Equation (8)). Upon the reaction coordinate mapping, a collective degree of freedom is included into the system, which in turn interacts with the residual environment, as sketched in (**c**) (see the main text for further details). For an underdamped $J_{SB}(\omega)$, $J_{RC}(\omega)$ adopts a Markovian form, as depicted in (**b**). Such interaction with a collective coordinate can be exploited to realize Hamiltonians containing multiphoton interaction terms, as indicated in (**c**) and explained in detail in Section 3. For structured environments, one can still rearrange the original environment using more collective coordinates into the augmented system S', where each of them interacts now with its own residual environment, as sketched in (**d**) (see Section 3.2 for further details).

3. Analogue Simulation of Multiphoton Spin-Boson Interactions

The task now consists of bringing the spin-boson Hamiltonian H_{SB} into the form of a n-photon model, i.e, into a model containing interaction terms of the form $\sigma^{\pm}a^n$ and $\sigma^{\pm}(a^\dagger)^n$. For that, one could perform the approximate mapping used in [43,44] directly onto H_{SB}. This would require the selection of a particular bosonic mode out of the environment with frequency ω_q to now play the role of a in the interaction with the spin ($c_q \to a$), while treating the rest of $c_{k\neq q}$ as a residual environment. Here, however, we resort to a more sophisticated procedure, based on the so-called reaction coordinate (RC) mapping [22–28], which consists of rearranging the environment degrees of freedom, such that a small number of collective coordinates can be included in the Hamiltonian part, which in turn interact with the residual environment. In certain cases, the open-quantum system description of the augmented system is considerably simplified with respect to the original system plus environment. Clearly, if the spin-boson model involves just a discrete number of modes, the reaction-coordinate procedure then trivially retrieves the original discrete environment.

3.1. Reaction Coordinate Mapping

In the following, we summarize how to make use of the RC mapping for a spin-boson model, which has been studied previously in different works [24,25], while referring to Appendix A and References [22–28] for further details of the calculations and of the RC mapping.

We shall start by defining a collective mode or reaction coordinate, described by the annihilation and creation operators a and $a^\dagger$, such that:

$$\lambda(a + a^\dagger) = \sum_k f_k(c_k + c_k^\dagger), \tag{6}$$

while the residual environmental degrees of freedom transform into b_k and $b_k^\dagger$, requiring that the latter appear in a normal form in the Hamiltonian. In this manner, the original spin-boson Hamiltonian adopts the form of $H_{\rm SB} = H_{\rm S+RC} + H_{\rm RC-E'} + H_{\rm E'}$, where the former is given by:

$$H_{\rm S+RC} = \frac{\Delta_0}{2}\sigma_x + \Omega a^\dagger a + \lambda\sigma_x(a + a^\dagger) + \sum_{j=0}^{n_d} \frac{\epsilon_j}{2}\left[\cos(\Delta_j - \Delta_0)t\,\sigma_z + \sin(\Delta_j - \Delta_0)t\,\sigma_y\right], \tag{7}$$

and the other two terms are $H_{\rm RC-E'} + H_{\rm E'} = (a + a^\dagger)\sum_k g_k(b_k + b_k^\dagger) + (a + a^\dagger)^2\sum_k \frac{g_k^2}{\omega_k} + \sum_k \omega_k b_k^\dagger b_k$. The reaction coordinate map is completed upon the identification of the parameters λ, Ω, and g_k or, thus, $J_{\rm RC}(\omega) = \sum_k g_k^2\delta(\omega - \omega_k)$. For certain cases, such mapping allow for an exact relation between the original and transformed parameters [28]. Indeed, considering an underdamped spin-boson spectral density in the initial spin-boson model,

$$J_{\rm SB}(\omega) = \frac{\alpha\Gamma\omega_0^2\omega}{(\omega_0^2 - \omega^2)^2 + \Gamma^2\omega^2}, \tag{8}$$

one can show that the resulting spectral density for the residual environment interacting with the reaction coordinate reads as:

$$J_{\rm RC}(\omega) = \gamma\omega e^{-\omega/\Lambda} \tag{9}$$

provided $\Lambda/\omega \gg 1$ and where the parameters are related according to $\gamma = \Gamma/(2\pi\omega_0)$, $\Omega = \omega_0$, and $\lambda = \sqrt{\pi\alpha\omega_0/2}$ (see Appendix A or [22–24,28] for further details of this derivation). Here, the frequency ω_0 in $J_{\rm SB}(\omega)$ denotes the position at which the spectral density features a maximum, while Γ and α account for its width and strength, respectively. For $J_{\rm RC}(\omega)$, the coupling strength is given by γ. In this manner, by augmenting the system incorporating a collective mode, the original spin-boson model with $J_{\rm SB}(\omega)$ is transformed into a spin plus reaction coordinate, which now in turn interacts with a Markovian environment, where the standard Born-Markov approximations can be performed [45]. Indeed, the master equation governing the dynamics of the augmented system, spin plus reaction coordinate, reads as (see Appendix A for the details of the calculation, which closely follows [24]):

$$\dot\rho_{\rm S+RC}(t) = -i\left[H_{\rm S+RC}, \rho_{\rm S+RC}(t)\right] - [x, [\chi, \rho_{\rm S+RC}(t)]] + [x, \{\Theta, \rho_{\rm S+RC}(t)\}]. \tag{10}$$

with $x = a + a^\dagger$, while the quantities χ and Θ define the rates affecting the reaction coordinate. They are defined as:

$$\chi \approx \frac{\pi}{2}\sum_{jk} J_{\rm RC}(\xi_{jk})\coth(\beta\xi_{jk}/2)x_{jk}\left|\phi_j\right\rangle\left\langle\phi_k\right|, \tag{11}$$

$$\Theta \approx \frac{\pi}{2}\sum_{jk} J_{\rm RC}(\xi_{jk})x_{jk}\left|\phi_j\right\rangle\left\langle\phi_k\right|, \tag{12}$$

where $x_{jk} = \left\langle\phi_j\right| x \left|\phi_k\right\rangle$, $H_{\rm S+RC}\left|\phi_j\right\rangle = \varphi_j\left|\phi_j\right\rangle$, and $\xi_{jk} = \varphi_j - \varphi_k$.

Having obtained the reaction coordinate Hamiltonian, we undertake the transformation of $H_{\rm S+RC}$, and thus, of Equation (10), to achieve a model that comprises spin-multiphoton interaction terms. For that purpose, we will introduce two auxiliary Hamiltonians H_a and H_b, which will arise in the intermediate steps by moving into a suitable interaction picture and transforming them accordingly. The first step consists indeed of moving to a rotating frame in which $H_{\rm S+RC} \equiv H^I_{a,1}$ where $H_a = H_{a,0} + H_{a,1}$ with $H_{a,0} = -\Delta_0/2\sigma_x$. In this manner, we find:

$$H_a = \Omega a^\dagger a + \lambda\sigma_x(a+a^\dagger) + \sum_{j=0}^{n_d} \frac{\epsilon_j}{2}\left[\cos\Delta_j t\sigma_z + \sin\Delta_j t\sigma_y\right]. \tag{13}$$

while Equation (10) transforms into:

$$\dot{\rho}_a(t) = -i\,[H_a, \rho_a(t)] - [x, [\hat{\chi}, \rho_a(t)]] + [x, \{\hat{\Theta}, \rho_a(t)\}]. \tag{14}$$

where $\hat{\chi} = U_{a,0}\chi U^\dagger_{a,0}$ and $\hat{\Theta} = U_{a,0}\Theta U^\dagger_{a,0}$, such that $U_x = \mathcal{T}e^{-i\int_0^t ds H_x(s)}$ is the time-evolution operator of a Hamiltonian H_x. Then, we perform a further transformation using the unitary operator $T(\alpha)$, defined as $T(\alpha) = 1/\sqrt{2}\left[D^\dagger(\alpha)\left(|e\rangle\langle e| - |g\rangle\langle e|\right) + D(\alpha)\left(|g\rangle\langle g| + |e\rangle\langle g|\right)\right]$ with $D(\alpha) = e^{\alpha a^\dagger - \alpha^* a}$ the standard displacement operator. Hence, $H_b \equiv T^\dagger(-\lambda/\Omega)H_a T(-\lambda/\Omega)$ such that $\rho_b = T^\dagger\rho_a T$, which leads to (see Appendix B for further details):

$$\dot{\rho}_b = -i\,[H_b, \rho_b] - \left[T^\dagger x T, \left[T^\dagger\hat{\chi}T, \rho_b(t)\right]\right] + \left[T^\dagger x T, \left\{T^\dagger\hat{\Theta}T, \rho_b(t)\right\}\right], \tag{15}$$

where the Hamiltonian H_b can be written as:

$$H_b = \Omega a^\dagger a + \sum_{j=0}^{n_d}\frac{\epsilon_j}{2}\left[\sigma^+ e^{2\lambda(a-a^\dagger)/\Omega}e^{-i\Delta_j t} + \text{H.c.}\right]. \tag{16}$$

Hence, the dissipator acting on ρ_b has the same form as in Equation (14), but with transformed operators, namely $T^\dagger x T$, $T^\dagger\hat{\chi}T$, and $T^\dagger\hat{\Theta}T$, where $T \equiv T(-\lambda/\Omega)$. Finally, by moving to an interaction picture with respect to $H_{b,0} = (\Omega - \tilde{\nu})a^\dagger a - \tilde{\omega}\sigma_z/2$ and expanding the exponential in Equation (16) (the latter requires that $|2\lambda/\Omega|\sqrt{\langle(a+a^\dagger)^2\rangle} \ll 1$ for truncating the exponential to a finite number of terms), we arrive at a Hamiltonian containing multiphoton interaction terms. The latter condition is commonly known as the Lamb-Dicke regime. In addition, we consider the driving frequencies to be $\Delta_j = \pm n_j(\tilde{\nu} - \Omega) - \tilde{\omega}$ with $|\Omega - \tilde{\nu}| \gg \epsilon_j/2$, so that one can safely perform a rotating-wave approximation keeping only those terms that are resonant, i.e., time independent (see Appendix B for further details of the calculation). Note that, as H_b is similar to the Hamiltonian describing an optical trapped ion under the action of lasers driving vibrational sidebands [48], the procedure to obtain Jaynes-Cummings or quantum Rabi models is analogous to those cases [35,49,50]. In this manner, we can approximate $H^I_{b,1} \equiv U^\dagger_{b,0}H_{b,1}U_{b,0} \approx H_{\rm n}$, where $H_{\rm n}$ contains the aimed at multiphoton interactions,

$$H_{\rm n} = \frac{\tilde{\omega}}{2}\sigma_z + \tilde{\nu}a^\dagger a + \sum_{j\in r}\frac{\epsilon_j(2\lambda)^{n_j}}{2\Omega^{n_j}n_j!}\left[\sigma^+ a^{n_j} + \text{H.c.}\right] + \sum_{j\in b}\frac{\epsilon_j(2\lambda)^{n_j}}{2\Omega^{n_j}n_j!}\left[\sigma^+(-a^\dagger)^{n_j} + \text{H.c.}\right]. \tag{17}$$

Note that the sets r and b encompass the terms with amplitude ϵ_j driving red- and blue-sidebands, that is those terms in Equation (5) with frequency $\Delta_{j\in r} = +n_j(\tilde{\nu} - \Omega) - \tilde{\omega}$ and $\Delta_{j\in b} = -n_j(\tilde{\nu} - \Omega) - \tilde{\omega}$. Each of these drivings will contribute with a multiphoton interaction, either $\sigma^+ a^{n_j} + \text{H.c.}$ for $j \in r$ or $\sigma^- a^{n_j} + \text{H.c.}$ for $j \in b$, which produce transitions between the states $|m\rangle|g\rangle \leftrightarrow |m \mp n_j\rangle|e\rangle$. We stress that for a time-independent spin-boson model, as given in Equations (1)–(4) (or equivalently with $n_d = 0$ in $H_{S,d}$ as given in Equation (5), one obtains a single n-photon [anti]-Jaynes-Cummings interaction term, $\sigma^+ a^n + \text{H.c.}$ $[\sigma^+(-a^\dagger)^n + \text{H.c.}]$, by choosing $\Delta_0 = n(\tilde{\nu} - \Omega) - \tilde{\omega}$ $[\Delta_0 = -n(\tilde{\nu} - \Omega) - \tilde{\omega}]$ in the original spin-boson Hamiltonian H_{SB}. Thus, one needs the knowledge of the relevant bosonic frequency Ω to simulate multiphoton interaction terms properly.

In order to show how the dissipative part transforms, it is advisable to introduce the time-dependent unitary operator:

$$\Phi = U_{b,0}^\dagger T^\dagger U_{a,0}. \tag{18}$$

Then, one can see that, defining $\tilde{\chi} = \Phi\chi\Phi^\dagger$, $\tilde{\Theta} = \Phi\Theta\Phi^\dagger$ and $\tilde{x} = \Phi(a + a^\dagger)\Phi^\dagger$, the resulting master equation for $\rho_{\text{n}}(t)$ is:

$$\dot{\rho}_{\text{n}}(t) = -i[H_{\text{n}}, \rho_{\text{n}}(t)] - [\tilde{x}, [\tilde{\chi}, \rho_{\text{n}}(t)]] + [\tilde{x}, \{\tilde{\Theta}, \rho_{\text{n}}(t)\}] \tag{19}$$

where the state $\rho_{\text{n}}(t)$ of the multiphoton model is related to the original spin-boson upon the reaction coordinate mapping, $\rho_{\text{S+RC}}(t)$, through a unitary transformation:

$$\rho_{\text{n}}(t) \approx \Phi\rho_{\text{S+RC}}(t)\Phi^\dagger. \tag{20}$$

From the previous expression, it follows that the purity of the total state $\rho_{\text{S+RC}}$ and that of ρ_{n} are approximately equal. Moreover, the reduced spin state in the different frameworks are related according to $\text{Tr}_{\text{B}}[\rho_{\text{SB}}(t)] = \text{Tr}_{\text{RC}}[\rho_{\text{S+RC}}(t)] \approx \text{Tr}_{\text{RC}}[\Phi^\dagger \rho_{\text{n}}(t)\Phi]$, where $\text{Tr}_{\text{B}}[\cdot]$ and $\text{Tr}_{\text{RC}}[\cdot]$ denote the trace over the environment degrees of freedom and reaction coordinate, respectively. In this manner, having access to the spin degree of freedom, one can have access to the dissipative spin dynamics dictated by the master Equation (19) under a multiphoton Hamiltonian H_{n}, given in Equation (17), whose parameters can be tuned. In addition, we remark that the initial state at $t_0 = 0$ in the multiphoton frame is related to that of the spin-boson model as $\rho_{\text{n}}(0) = T^\dagger \rho_{\text{S+RC}}(0) T$.

At this stage, a few comments regarding the validity of Equation (20) are in order. While the steps performed from $H_{\text{S+RC}}$ to H_b are exact, H_{n} is attained in an approximate manner. The good functioning of the simulation depends on how these approximations are met. That is, Equation (20) holds within the Lamb-Dicke regime $|2\lambda/\Omega|\sqrt{\langle (a + a^\dagger)^2\rangle} \ll 1$ and for parameters satisfying $|\Omega - \tilde{\nu}| \gg \epsilon_j/2\,\forall j$, so that one can perform a rotating-wave approximation. As a consequence, this approximation also sets a constraint on the total duration for a good simulation (see Appendix B). Note that, as the parameters λ and Ω are directly related to the original spin-boson spectral density, these conditions set constraints onto the accessible parameters, as well as on the temperature of the environment. Furthermore, in order to observe coherent multiphoton dynamics, the noise rates in Equation (19) must be small compared to the parameters involved in H_{n}. For the considered shape of $J_{\text{SB}}(\omega)$, this translates into $\Gamma \ll \tilde{\nu}, \tilde{g}_n$, where $\tilde{g}_n = \epsilon_0(2\lambda)^n/(2\Omega^n n!)$ for an $n_d = 0$ and $\Delta_0 = \pm n(\tilde{\nu} - \Omega) - \tilde{\omega}$ (cf. Equation (17).

Finally, we comment that the previous scheme can be carried out beyond the Lamb-Dicke regime [44]. Admittedly, when the Lamb-Dicke approximation does not hold, the Hamiltonian H_{n} is no longer a good approximation to the dynamics. In this case, the Hamiltonian H_{n} must be replaced by a suitable nonlinear Jaynes-Cummings or quantum Rabi model, whose coupling constants crucially depend on the Fock-state occupation number in a nonlinear fashion [51–54]. These nonlinear, yet multiphoton Hamiltonians appear then as a good approximation to H_b, and thus to H_{SB} whenever $|2\lambda/\Omega|\sqrt{\langle (a + a^\dagger)^2\rangle} \ll 1$ is not fulfilled, as recently shown in [44]. In this article, however, we will constrain ourselves to parameters within the Lamb-Dicke regime.

3.2. Structured Environments

As previously mentioned, the simulation of multiphoton spin-boson interactions is not restricted to a determined form of $J_{\rm SB}(\omega)$. Here, we show the derivation of the procedure to obtain an effective multiphoton Hamiltonian when the initial spin-boson model features a more complicated interaction with the environment. For simplicity, we consider that $J_{\rm SB}(\omega)$ can be split in two parts, $J_{\rm SB}(\omega) = J_{\rm SB,1}(\omega) + J_{\rm SB,2}(\omega)$, although its generalization to more is straightforward. The first contribution, $J_{\rm SB,1}(\omega)$, is considered here to be suitable for the realization of multiphoton interactions as described in Section 3.1. In addition, we will work under the assumption that the environment degrees of freedom corresponding to $J_{\rm SB,2}(\omega)$ can be treated and simplified using again a collective or reaction coordinate, as sketched in Figure 1c.

As discussed previously, we identify a collective coordinate for each of the contributions to the spectral density $J_{\rm SB}(\omega)$. In this manner, we augment the system to include both reaction coordinates, denoted here by $\mathrm{S}' = \mathrm{S} + \mathrm{RC}_1 + \mathrm{RC}_2$. Hence, its Hamiltonian is given by:

$$H_{\mathrm{S}'} = H_{\mathrm{S},d} + \Omega_1 a_1^\dagger a_1 + \lambda_1 \sigma_x (a_1 + a_1^\dagger) + \Omega_2 a_2^\dagger a_2 + \lambda_2 \sigma_x (a_2 + a_2^\dagger), \tag{21}$$

where $H_{\mathrm{S,d}}$ is the original spin Hamiltonian, which may contain spin rotations, introduced in Equation (5), while the subscripts denote the corresponding reaction coordinate. The parameters λ_i and Ω_i are determined by the spectral density $J_{\rm SB,i}(\omega)$. The dynamics of the augmented system S' is governed by the following master equation:

$$\begin{aligned}\dot{\rho}_{\mathrm{S}'}(t) = -i\left[H_{\mathrm{S}'}, \rho_{\mathrm{S}'}(t)\right] - \left[x_1, \left[\chi_1, \rho_{\mathrm{S}'}(t)\right]\right] - \left[x_2, \left[\chi_2, \rho_{\mathrm{S}'}(t)\right]\right] \\ + \left[x_1, \{\Theta_1, \rho_{\mathrm{S}'}(t)\}\right] + \left[x_2, \{\Theta_2, \rho_{\mathrm{S}'}(t)\}\right],\end{aligned} \tag{22}$$

where $x_i = a_i + a_i^\dagger$ for $i = 1, 2$, and χ_i and Θ_i are defined in analogy to Equations (11) and (12).

In order to find a suitable transformation to realize multiphoton interaction terms from $H_{\mathrm{S}'}$, we proceed in a similar manner as for a single reaction coordinate. That is, we first move to a rotating frame where $H_{\mathrm{S}'} \equiv H_{a,1}^I$, with $H_a = H_{a,0} + H_{a,1}$ and $H_{a,0} = -\Delta_0/2\sigma_x$. Therefore, the transformed Hamiltonian reads as:

$$H_a = \sum_{k=1,2} \Omega_k a_k^\dagger a_k + \lambda_k \sigma_x (a_k + a_k^\dagger) + \sum_j \frac{\epsilon_j}{2}\left[\cos \Delta_j t \sigma_z + \sin \Delta_j t \sigma_y\right]. \tag{23}$$

The next step is to perform the transformation using the unitary operator $T(\alpha)$. As previously mentioned, we consider that the first reaction coordinate is suitable for the quantum simulation of multiphoton interaction terms, due to the form of its spectral density. This argument enables one to choose $\alpha \equiv -\lambda_1/\Omega_1$, hence $H_b \equiv T^\dagger(-\lambda_1/\Omega_1) H_a T(-\lambda_1/\Omega_1)$. This transformation acts trivially on the second reaction coordinate, but it does affect the coupling between the latter and the spin. Finally, if we move to an interaction picture with respect to $H_{b,0} = (\Omega_1 - \tilde{\nu}_1) a_1^\dagger a_1 - \tilde{\omega}\sigma_z/2$, we obtain the Hamiltonian $H_{\mathrm{n},2} \approx H_{b,1}^I \equiv U_{b,0}^\dagger H_{b,1} U_{b,0}$,

$$\begin{aligned}H_{\mathrm{n},2} = \frac{\tilde{\omega}}{2}\sigma_z + \tilde{\nu} a_1^\dagger a_1 + \Omega_2 a_2^\dagger a_2 - \lambda_2 \sigma_z (a_2 + a_2^\dagger) + \sum_{j \in r} \frac{\epsilon_j}{2 n_j!}\left(\frac{2\lambda_1}{\Omega_1}\right)^{n_j}\left[\sigma^+ a_1^{n_j} + \mathrm{H.c.}\right] \\ + \sum_{j \in b} \frac{\epsilon_j}{2 n_j!}\left(\frac{2\lambda_1}{\Omega_1}\right)^{n_j}\left[\sigma^+ (-a_1^\dagger)^{n_j} + \mathrm{H.c.}\right],\end{aligned} \tag{24}$$

where we have considered $\Delta_j = \pm n_j(\tilde{\nu} - \Omega_1) - \tilde{\omega}$ and assumed the Lamb-Dicke regime $|\lambda_1/\Omega_1|\sqrt{\langle(a_1 + a_1^\dagger)^2\rangle} \ll 1$, and $|\Omega_1 - \tilde{\nu}| \gg \epsilon_j/2$ to perform a rotating-wave approximation. Note that, while the multiphoton terms are identical to those of $H_{\rm n}$ in Equation (17), the second reaction coordinate interacts with the spin degree of freedom. Indeed, depending on the parameters of $H_{\rm n,2}$, the effect of such an interaction may turn effectively into non-Markovian effects for the reduced state of the spin and first reaction coordinate, $\rho_{\rm n} = {\rm Tr}_2[\rho_{\rm n,2}]$. The final master equation governing the dynamics of $\rho_{\rm n,2}$ is:

$$\dot{\rho}_{\rm n,2}(t) = -i[H_{\rm n,2}, \rho_{\rm n,2}(t)] - [\tilde{x}_1, [\tilde{\chi}_1, \rho_{\rm n,2}(t)]] - [\tilde{x}_2, [\tilde{\chi}_2, \rho_{\rm n,2}(t)]] + [\tilde{x}_1, \{\tilde{\Theta}_1, \rho_{\rm n,2}(t)\}] + [\tilde{x}_2, \{\tilde{\Theta}_2, \rho_{\rm n,2}(t)\}] \quad (25)$$

where the operators involved are defined as in the case involving a single reaction coordinate (cf. Equation (19)). It is worth stressing that the relation between the states given in Equation (20) still holds. From the previous derivation, one can observe that the extension to more collective coordinates is straightforward.

4. Examples and Numerical Simulations

In this section, we provide examples of the previously-explained general theoretical framework to investigate the performance of the quantum simulation of different multiphoton Hamiltonians $H_{\rm n}$, as well as to discuss the limitation in the parameter regime for their realization. In particular, in Section 4.1, we first consider the case in which the original spin-boson model interacts just with a discrete number of modes, which can be viewed as a limit of vanishing spectral broadening $\Gamma \to 0$. This scenario will allow us to examine the validity of the required approximations without the effect of dissipation. Then, in Section 4.2, we will consider $\Gamma \neq 0$, where the reaction-coordinate mapping appears as a key step to realize a desired multiphoton Jaynes-Cummings model. The dynamics of each model is obtained by a standard numerical integration (fourth-order Runge-Kutta) of the corresponding master equation, namely Equations (10) and (19) for the spin-boson and multiphoton Jaynes-Cummings model, respectively. Note that for a structured environment, the master equations are given in Equations (22) and (25).

In all cases, we assess the performance of the realization of the targeted multiphoton Jaynes-Cummings models by means of the fidelity $F(t)$ between two states,

$$F(t) = {\rm Tr}\left[\sqrt{\sqrt{\rho_1(t)}\rho_2(t)\sqrt{\rho_1(t)}}\right]^2. \quad (26)$$

In particular, we will analyse to what extent is the relation given in Equation (20) satisfied. In other words, we will compare the aimed state of a multiphoton Jaynes-Cummings model $\rho_{\rm n}(t)$ with the one retrieved using the analogue simulator, $\Phi\rho_{\rm S+RC}(t)\Phi^\dagger$, that is $\rho_1(t) \to \rho_{\rm n}(t)$ and $\rho_2(t) \to \Phi\rho_{\rm S+RC}(t)\Phi^\dagger$ in Equation (26). We remark that when two reaction coordinates are included, the state $\rho_{\rm n}(t)$ obeys the master equation given in Equation (25), whose Hamiltonian is $H_{\rm n,2}$, Equation (24), while $\rho_{\rm S+RC}(t)$ must be replaced by $\rho_{S'}$, as explained in Section 3.2.

In addition, we will show that the theoretical framework allows us to realize non-Markovian multiphoton Jaynes-Cummings models. Among the different measures for non-Markovianity [55], we resort to the one based on the trace distance [56], defined as:

$$\mathcal{D}(\rho_x, \rho_y) = \frac{1}{2}{\rm Tr}\left[|\rho_x - \rho_y|\right]. \quad (27)$$

where $|A| = \sqrt{A^{\dagger}A}$. Then, non-Markovian evolutions can be characterized as those for which $\mathcal{D}(\rho_x(t), \rho_y(t))$ increases during certain time intervals, that is for those for which the time-derivative of the trace distance for a pair of states $\rho_{x,y}$,

$$\sigma(t, \rho_{x,y}) = \frac{d}{dt}\mathcal{D}(\rho_x(t), \rho_y(t)), \tag{28}$$

is $\sigma(t, \rho_{x,y}) > 0$. In general, one has to maximize over all possible pairs of states $\rho_{x,y}$ in order to find a suitable non-Markovian measure [56]. For our purpose, however, it will be sufficient to show that $\sigma(t, \rho_{x,y}) > 0$ for a certain pair of states in a multiphoton Jaynes-Cummings model and that it can be retrieved using a spin-boson model. That is, we calculate $\sigma(t, \rho_{x,y})$ using two initial states $\rho_{x,y}$ in the multiphoton Jaynes-Cummings model and corroborate that $\sigma(t, \rho_{x,y})$ is obtained to a very good approximation when the states $\rho_{x,y}(t)$ are replaced by their simulated ones using the spin-boson model, namely $\rho_x(t) \to \Phi\rho_{x,\mathrm{S+RC}}(t)\Phi^{\dagger}$ and $\rho_y(t) \to \Phi\rho_{y,\mathrm{S+RC}}(t)\Phi^{\dagger}$. In this manner, we offer a proof-of-principle that non-Markovian multiphoton models can be realized.

4.1. Dissipationless Multiphoton Jaynes-Cummings Models

We start considering the simplest case, namely when the spin-boson model simply involves the interaction with a discrete number of modes. This corresponds to either considering $\Gamma \to 0$ in the underdamped spectral density $J_{\mathrm{SB}}(\omega)$ or, equivalently, assuming that dissipation effects are sufficiently small so that they can be discarded. Note that for a single bosonic mode with $\Gamma = 0$, the spin-boson model adopts the form of a generalized quantum Rabi model, which is indeed $H_{\mathrm{S+RC}}$, as given in Equation (7). Recall that in this particular case, $H_{\mathrm{SB}} \equiv H_{\mathrm{S+RC}}$, as there are no further modes in the system. In particular, we set $n_d = 0$ in Equation (5) as we aim to realize a single multiphoton Jaynes-Cummings interaction. The Hamiltonian for a nJCM can be written in general as:

$$H_{\mathrm{nJCM}} = \frac{\tilde{\omega}}{2}\sigma_z + \tilde{\nu}a^{\dagger}a + \tilde{g}_n\left(\sigma^{+}a^{n} + \sigma^{-}(a^{\dagger})^{n}\right). \tag{29}$$

At resonant condition, $\tilde{\omega} = n\tilde{\nu}$, the coupling constant $\tilde{g}_n$ fixes the time required to transfer the population from the state $|e\rangle|0\rangle$ to $|g\rangle|n\rangle$, denoted as $\tau_n = \pi/(2\tilde{g}_n\sqrt{n!})$. Both are related to the spin-boson parameters as (cf. Equation (17)):

$$\tilde{g}_n = \frac{\epsilon_0}{2\,n!}\left(\frac{2\lambda}{\Omega}\right)^{n} \tag{30}$$

$$\tau_n = \frac{\sqrt{n!}}{\epsilon_0}\left(\frac{\Omega}{2\lambda}\right)^{n}. \tag{31}$$

Clearly, as $2\lambda/\Omega$ must be small to lie within the Lamb-Dicke regime, the coupling $\tilde{g}_n$ decreases considerably for increasing n, requiring longer evolution times under the spin-boson Hamiltonian to observe a significant effect, that is an evolution time of the order of τ_n.

In Figure 2, we show the results for the realization of 2JCM and 3JCM models using a spin-boson model interacting with a single bosonic mode. In order to observe the paradigmatic Rabi oscillations between the states $|e\rangle|0\rangle$ and $|g\rangle|n\rangle$, we choose $\rho_{\mathrm{S+RC}}(0) = |-\rangle\langle-| \otimes \rho_{\mathrm{RC}}^{\mathrm{th}}$ as an initial state for the spin-boson model, where $\rho_{\mathrm{RC}}^{\mathrm{th}}$ is a thermal state at temperature β^{-1} for the reaction coordinate mode, containing $n^{\mathrm{th}} = (e^{\beta\Omega} - 1)^{-1}$ bosons. Recall that, as we consider here a single spectral density with $\Gamma = 0$, the reaction coordinate mode is simply the unique mode that interacts with the spin degree of freedom. In this manner, the initial state for the simulated multiphoton models reads as $\rho_{\mathrm{nJCM}}(0) = T^{\dagger}\rho_{\mathrm{S+RC}}(0)T$, which approximately amounts to $\rho_{\mathrm{nJCM}}(0) \approx |e\rangle\langle e| \otimes |0\rangle\langle 0|$ for sufficiently low temperature and small

$2\lambda/\Omega$. The chosen parameters for the simulation of the 2JCM, plotted in Figure 2a,b, are $\pi\alpha = \epsilon_0 = 0.02\omega_0$; recalling that $\Omega = \omega_0$, it results in $2\lambda/\Omega = 0.2$. Choosing $\tilde{\nu} = 10^{-3}\Omega$ and $\tilde{\omega} = 2\tilde{\nu}$, the coupling in 2JCM amounts to $\tilde{g}_2 = 0.2\tilde{\nu}$. The initial reaction-coordinate thermal state, $\rho_{\rm RC}^{\rm th}$, contains $n^{\rm th} = 10^{-3}$ bosons. In Figure 2b, we show how the quantum simulation of the 2JCM model deteriorates for increasing number of bosons, as a large $n^{\rm th}$ will eventually break down the Lamb-Dicke regime.

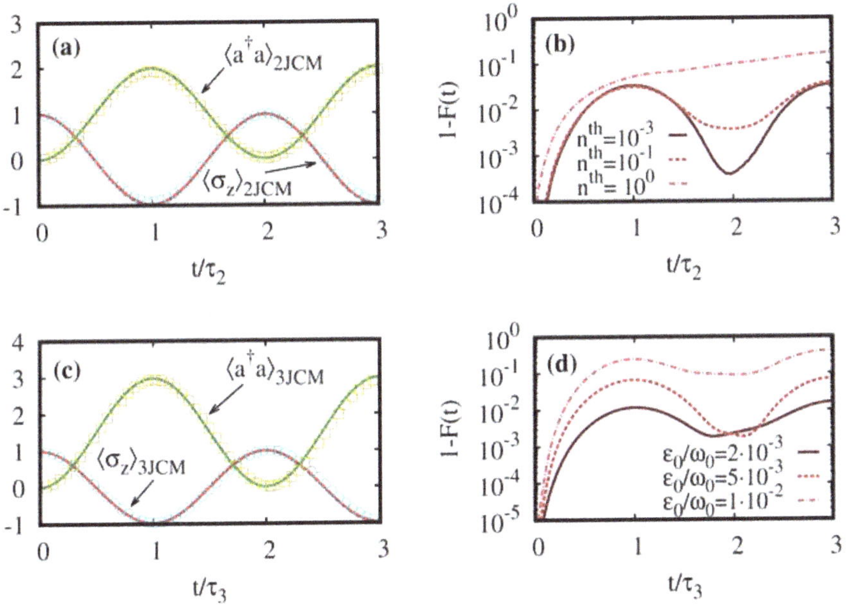

Figure 2. Dynamics of the simulated multiphoton Jaynes-Cummings models, $n = 2$ (top) and $n = 3$ (bottom). In Panels (**a**) and (**c**), we show the targeted dynamics (solid lines) and the one obtained using the spin-boson Hamiltonian (points) for $\langle a^\dagger a \rangle$ and $\langle \sigma_z \rangle$, as indicated in the plots and as a function of the time rescaled by τ_n (Equation (31)). In Panels (**b**) and (**d**), we plot the infidelity $1 - F(t)$ between the ideal $\rho_{\rm nJCM}(t)$ state and its approximated one $\Phi\rho_{\rm S+RC}(t)\Phi^\dagger$ for different conditions, namely in (**b**) for different temperatures (or mean occupation number $n^{\rm th}$) and in (**d**) for different values of ϵ_0/Ω. See Section 4.1 for further details regarding the parameters and states considered in the simulation. JCM, Jaynes-Cummings model.

For the 3JCM, we choose again $\pi\alpha = 0.02\omega_0$, which leads to $2\lambda/\Omega = 0.2$. Then, we select the aimed coupling strength of the multiphoton interaction to be $\tilde{g}_3 = 0.1\tilde{\nu}$ with $\tilde{\omega} = 3\tilde{\nu}$, while we vary ϵ_0/ω_0. The temperature is set to $\beta\Omega \approx 100$ so that $\rho_{\rm RC}^{\rm th} \approx |0\rangle\langle 0|$. As in the previous case, the dynamics are well retrieved; see Figure 2c, where we have set $\epsilon_0/\omega_0 = 2 \cdot 10^{-3}$. Note however that, as a consequence of the rotating-wave approximation performed to achieve a resonant third order (see Appendix B and cf. Equation (17)) and due to the longer times required to simulate a 3JCM compared to the 2JCM, the condition $|\Omega - \tilde{\nu}| \gg \epsilon_0$ must be better satisfied. Indeed, for $\epsilon_0/\omega_0 = 10^{-2}$, we already see a clear departure from the targeted dynamics, as indicated by a large infidelity $1 - F(t) \gtrsim 10^{-1}$, as shown in Figure 2d.

In the following, we consider a spin interacting with two bosonic modes, again with $\Gamma_{1,2} = 0$. As explained in Section 3.2, we perform the map onto the first bosonic mode to attain a multiphoton interaction. Upon suitable transformations and approximations, the spin-boson model will take the form of a multiphoton Jaynes-Cummings model $H_{\text{nJCM,2}}$, where the subscript 2 indicates the presence of a second reaction coordinate in the system. The Hamiltonian $H_{\text{nJCM,2}}$ reads as:

$$H_{\text{nJCM,2}} = \frac{\tilde{\omega}}{2}\sigma_z + \tilde{\nu}a_1^\dagger a_1 + \Omega_2 a_2^\dagger a_2 + \tilde{g}_n\left(\sigma^+ a_1^n + \sigma^-(a_1^\dagger)^n\right) - \lambda_2\sigma_z(a_2 + a_2^\dagger). \tag{32}$$

In this manner, the spin exchanges n quanta with the first bosonic mode as in H_{nJCM}, while the last term effectively shifts the spin frequency depending on the state of the second mode. The reduced state for the spin and first bosonic mode is given then by $\rho_{\text{nJCM}}(t) = \text{Tr}_2[\rho_{\text{nJCM,2}}(t)]$. Indeed, due to the interaction with the second bosonic mode, the multiphoton Jaynes-Cummings model may exhibit non-Markovian features. For that, we consider the spin-boson Hamiltonian $H_{S'}$ given in Equation (21), which then approximately realizes $H_{\text{nJCM,2}}$. In particular, we select $\Delta_0 = -2\Omega_1$, so that the simulated model involves two-photon interaction terms, i.e., a 2JCM. The results are plotted in Figure 3, while the parameters are $\pi\alpha_i = 0.02\Omega_i$ such that $2\lambda_i/\Omega_i = 0.2$ for $i = 1, 2$, $\epsilon_0/\Omega_1 = 10^{-2}$. The coupling strength in $H_{\text{2JCM,2}}$ is given by $\tilde{g}_2 = 0.2\tilde{\nu}$ with $\tilde{\nu} = \Omega_2$. As in the single-mode case, Rabi oscillations will be clearly visible selecting $\rho_{S'}(0) = |-\rangle\langle -| \otimes \rho^{\text{th}}_{\text{RC}_1} \otimes \rho^{\text{th}}_{\text{RC}_2}$. After its transformation, this state corresponds approximately to an initial spin state $|e\rangle$ in the nJCM frame. In the same manner, in order to analyse the emergence of non-Markovian behaviour, we consider the initial states $|g\rangle\langle g|$ and $|e\rangle\langle e|$ for the spin in $H_{S'}$. This implies initial spin states $|\pm\rangle$ in the nJCM frame, which for pure dephasing noise, it has been shown to be the pair of states maximizing $\sigma(t)$ [56]. The results plotted in Figure 3 have been performed considering a sufficiently low temperature such that $\rho^{\text{th}}_{\text{RC}_{1,2}} \approx |0\rangle\langle 0|$. We then compute the trace distance $\mathcal{D}(\rho_x, \rho_y)$ using the states $\rho_{x,y}(t)$ resulting in tracing out the second mode, $\text{Tr}_2[\rho_{\text{2JCM,2}}(t)]$, for the two different initial states $\rho_{2JCM,2}(0) \approx |\pm\rangle\langle\pm| \otimes \rho^{\text{th}}_{\text{RC}_1} \otimes \rho^{\text{th}}_{\text{RC}_2}$. As shown in Figure 3b, the time-derivative of the trace distance, $\sigma(t)$, becomes positive during certain intervals, a clear indication of the non-Markovian behaviour of the simulated multiphoton Jaynes-Cummings model. In addition, we also calculate the non-trivial evolution of the purity for the states $\rho_{\text{S+RC}_1}(t)$ and $\rho_{\text{S}}(t) = \text{Tr}_{\text{RC}_1}[\rho_{\text{S+RC}_1}(t)]$, which is shown in Figure 3c. According to our theoretical framework, their purity is approximately equal to that of $\rho_{\text{2JCM}}(t)$ and the reduced spin state upon tracing both bosonic degree of freedom in the 2JCM, $\text{Tr}[\rho_{\text{2JCM}}(t)]$, respectively. Finally, the infidelity $1 - F(t)$ between the targeted state $\rho_{\text{2JCM,2}}(t)$ and its reconstructed one $\Phi\rho_{\text{S+RC}_1+\text{RC}_2}(t)\Phi^\dagger$ in Figure 3d.

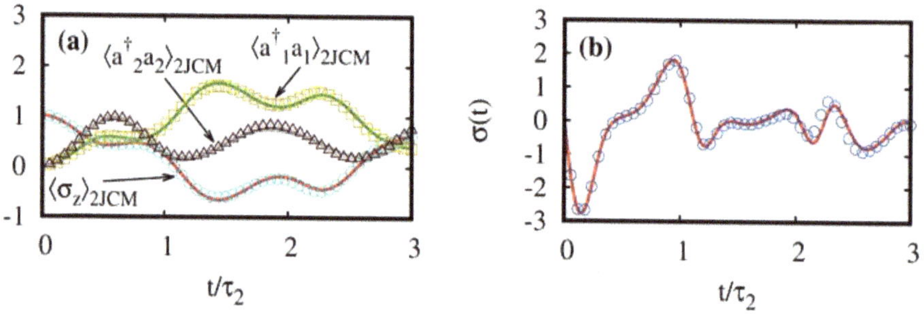

Figure 3. *Cont.*

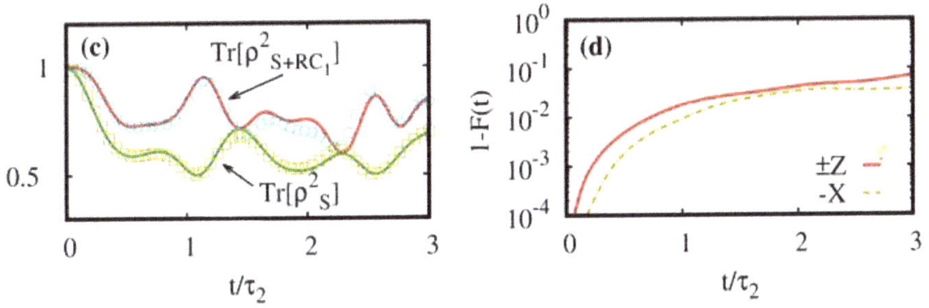

Figure 3. Non-Markovian dynamics for a 2JCM and its simulation using a spin-boson model $H_{S'}$. In Panel (**a**), we show the dynamics for the expectation values $\langle a_i^\dagger a_i\rangle$ with $i = 1,2$ and $\langle\sigma_z\rangle$ for the target 2JCM model (solid lines) and its reconstructed values using $H_{S'}$ (points). The considered initial state reads as $\rho_{S'}(0) = |-\rangle\langle -|\otimes\rho^{\rm th}_{\rm RC_1}\otimes\rho^{\rm th}_{\rm RC_2}$, with β very large such that $\rho^{\rm th}\approx|0\rangle\langle 0|$. In (**b**), we plot the time-derivative of the trace distance, $\sigma(t)$, after tracing out the second bosonic mode and considering the initial states $|e\rangle$ and $|g\rangle$ for the spin in $H_{S'}$, while both reaction coordinates find themselves in their vacuum. Clearly, $\sigma(t) > 0$ during certain intervals, revealing the non-Markovianity introduced due to the interaction with the second mode. Panel (**c**) shows the evolution of purity for the state upon tracing the second mode, $\mathrm{Tr}[\rho^2_{\rm S+RC_1}(t)]$ and for the reduced state of the spin, $\mathrm{Tr}[\rho^2_{\rm S}(t)]$, for the same case shown in (**a**). In Panel (**d**), we compare the infidelity $1 - F(t)$ between the ideal state and the simulated one using $H_{S'}$ for the three different initial states employed here. We refer to Section 4.1 for further details regarding the parameters and states considered in the simulation.

4.2. Dissipative Multiphoton Jaynes-Cummings Models

We now consider a more realistic scenario in which the spin-boson model interacts with an environment whose spectral density has an underdamped shape, i.e., $J_{\rm SB}(\omega)$ has the form of Equation (8) with $\Gamma\neq 0$. In this manner, we extend the theoretical framework beyond the standard local master equation description [44]. As explained in Section 3.1, this situation can be mapped using a reaction coordinate, which now in turn interacts with a Markovian residual environment. The evolution of the state of the augmented system, spin and reaction coordinate, evolves according to the master equation given in Equation (10). Indeed, the effect of spectral broadening, $\Gamma\neq 0$, introduces dissipation into the simulated multiphoton Jaynes-Cummings model, whose state now obeys the master Equation (19). We remark that the performance of the simulated dissipative model is not altered when the effect of dissipation is taken into account correctly. Nevertheless, whenever $\Gamma\gg\tilde{\nu}$, dissipation dominates the dynamics, and the paradigmatic Rabi oscillations will eventually fade away. In Figure 4, we show the results of numerical simulations aimed to retrieve a 2JCM with different $\Gamma/\tilde{\nu}$ values and for different quantities. As for Figure 2, we used $\pi\alpha = \epsilon_0 = 0.02\omega_0$, so that $2\lambda/\Omega = 0.2$. We chose again $\tilde{\nu} = 10^{-3}\Omega$ and $\tilde{\omega} = 2\tilde{\nu}$, and therefore, the coupling in 2JCM amounts to $\tilde{g}_2 = 0.2\tilde{\nu}$, while the temperature is such that $\rho^{\rm th}_{\rm RC}$ contains $n^{\rm th} = 10^{-3}$ bosons. The spin is initialized in the $|-\rangle$ state, so that $\rho_{\rm S+RC}(0) = |-\rangle\langle -|\otimes\rho^{\rm th}_{\rm RC}$. In particular, the value $\Gamma/\tilde{\nu} = 2\cdot 10^{-1}$ considered in Figure 4a already produces a significant departure from the Rabi oscillation between the states $|e\rangle|0\rangle$ and $|g\rangle|2\rangle$ in the dissipationless 2JCM (cf. Figure 2a for $\Gamma = 0$). Note that the results plotted in Figure 4a correspond to a critically-damped 2JCM since $\Gamma = \tilde{g}_2$. As plotted in Figure 4b, the effect of the dissipation is clearly visible in the evolution of the purity for both the total state (spin plus bosonic mode) and the reduced spin state, namely $\mathrm{Tr}[\rho^2_{\rm S+RC}(t)]$ and $\mathrm{Tr}[\rho^2_{\rm S}(t)]$. As in previous cases, the purity of these states is directly related to those of the simulated model as a consequence of the relation $\rho_{\rm 2JCM}(t)\approx\Phi\rho_{S+RC}(t)\Phi^\dagger$. Furthermore, Rabi oscillations or population revivals appear in the evolution of von Neumann entropy, $S_{\rm vN}(\rho) = -\rho\log_2\rho$ for the reduced spin state. In particular, for an initial state $\rho_{nJCM}(0) \approx |e\rangle\langle e| \otimes |0\rangle\langle 0|$ and due to the n-photon interaction with a bosonic degree of freedom, the spin state oscillates between a pure ($S_{\rm vN} = 0$) and a maximally-mixed state ($S_{\rm vN} = 1$) in a time $\tau_n/2$. This further corroborates that one can witness the multiphoton transitions of the aimed multiphoton Jaynes-Cummings model monitoring the spin even without access or control on the bosonic environment. This is plotted in Figure 4c for different $\Gamma/\tilde{\nu}$ values. Finally, we note that the performance of the quantum simulation is independent of the dissipation as demonstrated by the good fidelities attained in these cases (cf. Figure 4d), allowing for the simulation of different parameter regimes in a nJCM.

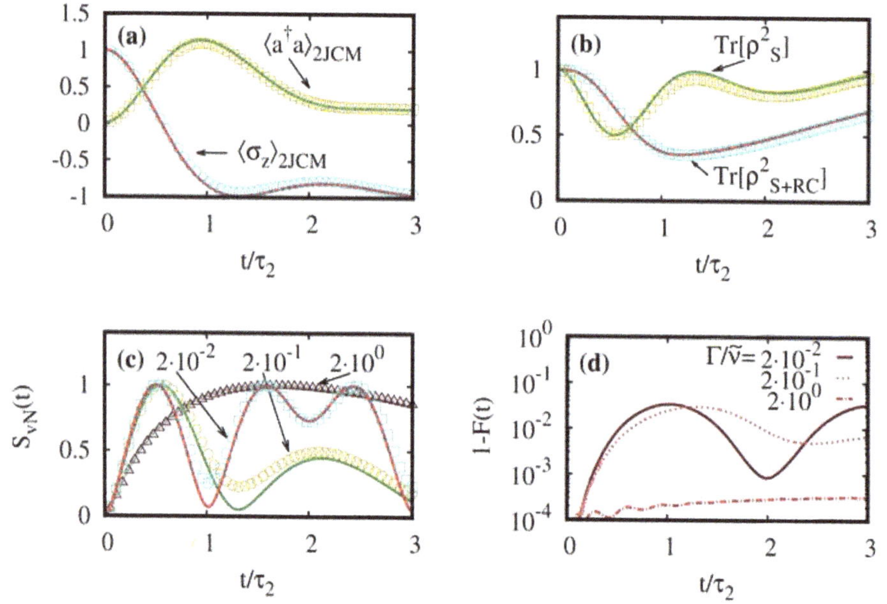

Figure 4. Dynamics of a dissipative 2JCM using a spin-boson model. In Panel (**a**), we show the dynamics of the expectation values of $\langle a^\dagger a\rangle$ and $\langle \sigma_z\rangle$, as in Figure 2, for the dissipative 2JCM (solid lines) and its simulation using the spin-boson model (points), for $\Gamma/\tilde{\nu} = 2\cdot 10^{-1}$ and $\rho_{\rm S+RC}(0) = |-\rangle\langle -| \otimes \rho_{\rm RC}^{\rm th}$ with $n^{\rm th} = 10^{-3}$. For the same case, we also show in (**b**) the evolution of the purities for the spin state ${\rm Tr}[\rho_{\rm S}^2(t)]$ and for the total state ${\rm Tr}[\rho_{\rm S+RC}^2(t)]$. In (**c**), we compare the different behaviour as $\Gamma/\tilde{\nu}$ varies for the von Neumann entropy of the reduced spin state, $S_{\rm vN}(\rho_{\rm S}(t))$. The values of $\Gamma/\tilde{\nu}$ are indicated close to each curve. Finally, the state infidelity $1 - F(t)$ between the targeted $\rho_{\rm 2JCM}$ and its approximate simulation, $\Phi\rho_{\rm S+RC}(t)\Phi^\dagger$, is plotted in Panel (**d**) for different $\Gamma/\tilde{\nu}$. See the main text for further details on the parameters employed for the simulation.

5. Conclusions

We have proposed a theoretical scheme to realize multiphoton Jaynes-Cummings models using the paradigmatic spin-boson model, which contains a continuum of bosonic modes, as an analogue quantum simulator. While the spin-boson model naturally lacks these multiphoton interaction terms, we make use of a suitable transformation that approximately maps the spin-boson model into a dissipative multiphoton Jaynes-Cummings model. Importantly, the parameters of the multiphoton model, as well as the order of the interaction can be controlled by tuning the frequency splitting and bias parameter of the spin in the original spin-boson model. In order to bring the spin-boson model, typically interacting with an infinite number of bosonic modes, into the form of the aimed multiphoton model, we first rearrange the environment degrees of freedom using the so-called reaction-coordinate method [22–28]. This method allows us to include a set of collective bosonic modes into the coherent description of the problem, which then in turn interact with the residual environment. For certain types of interactions between the spin and the environment, characterized by the spectral density, the reaction coordinate mapping emerges as a powerful tool to reduce the complexity of the problem. In particular, for an underdamped spectral density, the reaction coordinate takes a simple form as it interacts with the residual environment in a Markovian fashion. The resulting Hamiltonian is then used to generate multiphoton interaction terms, following the theory explained in [43,44], while the dissipation effects must be transformed accordingly. Furthermore, we extend the scheme to spin-boson models with structured environments. In these cases, the original spin-boson Hamiltonian can be mapped onto the one of a spin interacting with more reaction coordinates. In this manner, we show how to extend the theoretical framework to account for these additional modes. In particular, due to the presence of two or more reaction coordinates, the attained multiphoton Jaynes-Cummings model can exhibit non-Markovian features. We perform numerical simulations starting from the spin plus reaction-coordinate Hamiltonians and aiming to realize different multiphoton Jaynes-Cummings models. We first perform simulations considering one reaction coordinate without dissipation to better illustrate the performance of the required approximations to achieve two- and three-photon Jaynes-Cummings models. We then demonstrate that non-Markovian multiphoton Jaynes-Cummings models can be indeed attained when a second reaction coordinate is included, as unveiled by the standard trace distance measure [56]. Finally, we provide numerical simulations investigating the interplay between spectral broadening, dissipation and the decoherence in the targeted multiphoton models.

Author Contributions: Conceptualization R.P. and J.C.; investigation, R.P., G.Z., I.A., E.S., M.P. and J.C.; supervision, E.S., M.P. and J.C.; writing R.P.; writing, review and editing, G.Z., I.A., E.S., M.P. and J.C.

Funding: G.Z. is supported by the H2020-MSCA-COFUND-2016 project SPARK(Grant No. 754507). R.P. and M.P. acknowledge the support by the SFI-DfEInvestigator Programme (Grant 15/IA/2864). M.P. acknowledges the H2020 Collaborative Project TEQ(Grant Agreement 766900), the Leverhulme Trust Research Project Grant UltraQuTe (Grant No. RGP-2018-266) and the Royal Society Wolfson Fellowship (RSWF\R3\183013). J.C. acknowledges support by the Juan de la Cierva Grant IJCI-2016-29681. I.A. acknowledges support by Basque Government Ph.D. Grant No. PRE-2015-1-0394. We also acknowledge funding from Spanish MINECO/FEDER FIS2015-69983-P and Basque Government IT986-16. This material is also based on work supported by the U.S. Department of Energy, Office of Science, Office of Advance Scientific Computing Research (ASCR), Quantum Algorithm Teams (QAT) Program under Field Work Proposal Number ERKJ333. J.C. and E.S. acknowledge support from the projects QMiCS(820505) and OpenSuperQ(820363) of the EU Flagship on Quantum Technologies.

Conflicts of Interest: The authors declare no conflict of interest. The funders had no role in the design of the study; in the collection, analyses, or interpretation of data; in the writing of the manuscript; nor in the decision to publish the results.

Appendix A. Reaction Coordinate Mapping

In this Appendix, we provide the necessary steps for the reaction coordinate mapping, as well as for the derivation of the master equation given in Equation (10), following closely [24]. As outlined in Section 3.1, given the Hamiltonian of the spin-boson system $H_{\rm SB} = \frac{\epsilon_0}{2}\sigma_z + \frac{\Delta_0}{2}\sigma_x + \sigma_x \sum_k f_k(c_k + c_k^\dagger) + \sum_k \omega_k c_k^\dagger c_k$, one can achieve the RC mapping by defining a collective coordinate such that $\lambda(a + a^\dagger) = \sum_k f_k(c_k + c_k^\dagger)$, where a and $a^\dagger$ are respectively the annihilation and creation operators of the RC. This transformation leads to a new Hamiltonian where the original system interacts with the residual environment only through the RC,

$$H = H_{\rm S+RC} + H_{\rm RC-E'} + H_{\rm E'}, \tag{A1}$$

where $H_{\mathrm{S+RC}}$ is given by Equation (7), while $H_{\mathrm{RC-E'}} = (a+a^\dagger)\sum_k g_k(b_k+b_k^\dagger) + (a+a^\dagger)^2\sum_k \frac{g_k^2}{\omega_k}$, $H_{\mathrm{E'}} = \sum_k \omega_k b_k^\dagger b_k$.

The crucial point of this procedure is to find an explicit relation between the spectral density of the original configuration, i.e. $J_{\mathrm{SB}}(\omega) = \sum_k f_k^2\delta(\omega-\omega_k)$, and the analogue quantity of the transformed system $J_{\mathrm{RC}}(\omega) = \sum_k g_k^2\delta(\omega-\omega_k)$. In order to obtain this relation, one can rephrase the problem classically. Indeed, since the spectral density only depends on the interaction between the system and the environment, one can momentarily regard the spin as a continuous coordinate q subject to a potential $V(q)$. After solving the corresponding Hamilton equations of motion in the Fourier space, one obtains an equation of the form $\hat{L}_{\mathrm{SB}}(z)\hat{q}(z) = -\hat{V}'(z)$, where $\hat{L}_{\mathrm{SB}}(z) = -z^2\left(1+\int_0^{+\infty} d\omega \frac{2J_{\mathrm{SB}}(\omega)}{\omega(\omega^2-z^2)}\right)$. Therefore, using the so-called Leggett prescription, one gets:

$$J_{\mathrm{SB}}(\omega) = \frac{1}{\pi}\lim_{\epsilon\to 0^+} \mathrm{Im}\left[\hat{L}_{\mathrm{SB}}(\omega - i\epsilon)\right]. \tag{A2}$$

One can reproduce the same calculation also after performing the RC mapping and express $J_{\mathrm{RC}}(\omega)$ in terms of the corresponding kernel $\hat{L}_0(z)$. However, since at this stage, we are just rearranging the environment in a more convenient way by using a suitable normal mode transformation, the integral kernel must be the same before and after the mapping; hence, one can use $\hat{L}_0(z)$ instead of $\hat{L}_{\mathrm{SB}}(z)$ in Equation (A2). By considering the Ohmic spectral density $J_{\mathrm{RC}}(\omega) = \gamma\omega e^{-\omega/\Lambda}$, one obtains:

$$J_{\mathrm{SB}}(\omega) = \frac{4\gamma\Omega^2\lambda^2\omega}{(\Omega^2-\omega^2)^2 + (2\pi\gamma\Omega\omega)^2}. \tag{A3}$$

It is easy to see that one exactly recovers the underdamped spectral density given by Equation (8) by simply requiring that $\gamma = \Gamma/(2\pi\omega_0)$, $\Omega = \omega_0$, and $\lambda = \sqrt{\pi\alpha\omega_0/2}$. Furthermore, one also needs to solve the dynamics, i.e., writing down the corresponding master equation for the mapped system, system plus reaction coordinate. The guiding idea is to treat exactly the coupling between the spin and the RC, while the interaction between the latter and the residual environment is treated perturbatively up to the second order. This enables us to rely on the standard Born-Markov approximation, provided that either the coupling between the augmented system and the residual environment is weak or the residual environment correlation time is short compared to the relevant time scale of the system. Within this approximation, one can work out a master equation that, in the Schrödinger picture, reads as:

$$\begin{aligned}\dot{\rho}(t) = &-i\left[H_{\mathrm{S+RC}}, \rho(t)\right] - \int_0^\infty d\tau \int_0^\infty d\omega J_{\mathrm{RC}}(\omega)\cos\omega\tau \coth\left(\frac{\beta\omega}{2}\right)\left[A,\left[A(-\tau),\rho(t)\right]\right] \\ &- \int_0^\infty d\tau \int_0^\infty d\omega J_{\mathrm{RC}}(\omega)\frac{\cos\omega\tau}{\omega}\left[A,\left\{\left[A(-\tau), H_{\mathrm{S+RC}}\right],\rho(t)\right\}\right],\end{aligned}$$

where $\rho \equiv \rho_{\mathrm{S+RC}}$, $A = a + a^\dagger$, and the residual environment is assumed to be in a thermal state, i.e., $\rho_{E'} = e^{-\beta H_{\mathrm{E'}}}/\mathrm{Tr}_{\mathrm{E'}}\{e^{-\beta H_{\mathrm{E'}}}\}$.

In order to obtain an expression for the interaction picture operators, one can proceed by truncating the space of the augmented system up to n basis states and numerically diagonalising the Hamiltonian $H_{\mathrm{S+RC}}$. To this end, let $|\phi_n\rangle$ be an eigenstate of $H_{\mathrm{S+RC}}$, i.e., $H_{\mathrm{S+RC}}|\phi_j\rangle = \varphi_j|\phi_j\rangle$; therefore, the operator A can be expanded as $A = \sum_{jk} A_{jk}|\phi_j\rangle\langle\phi_k|$, while in the interaction picture, one has:

$$A(t) = \sum_{jk} A_{jk} e^{i\xi_{jk}t}|\phi_j\rangle\langle\phi_k|, \tag{A4}$$

where $A_{jk} = \langle \phi_j | A | \phi_k \rangle$, and $\xi_{jk} = \varphi_j - \varphi_k$. Finally, by plugging Equation (A4) into Equation (A4) and assuming the imaginary parts to be negligible, one gets the final form of the master equation given by Equation (10).

Appendix B. Derivation of H_b and $H_{\rm n}$

In this Appendix, we show how to obtain the Hamiltonians H_b and $H_{\rm n}$, given in Equations (16) and (17), respectively. In particular, for H_b, the following expressions are needed:

$$\begin{aligned} T^\dagger(\alpha) a^\dagger a T(\alpha) &= a^\dagger a + |\alpha|^2 - \sigma_z(a\alpha^* + a^\dagger \alpha), \\ T^\dagger(\alpha)\sigma_x T(\alpha) &= -\sigma_z, \\ T^\dagger(\alpha)\sigma_y T(\alpha) &= -iD(2\alpha)\sigma^+ + \text{H.c.}, \\ T^\dagger(\alpha)\sigma_z T(\alpha) &= D(2\alpha)\sigma^+ + \text{H.c.}, \\ T^\dagger(\alpha)\sigma_x(a+a^\dagger)T(\alpha) &= -\sigma_z(a+a^\dagger) + 2\text{Re}[\alpha]. \end{aligned}$$

Thus, the resulting Hamiltonian $H_b = T^\dagger H_a T$, with $H_a = \Omega a^\dagger a + \lambda \sigma_x (a + a^\dagger) + \sum_j \epsilon_j/2(\cos \Delta_j t \sigma_z + \sin \Delta_j t \sigma_y)$, reads:

$$H_b = = \Omega a^\dagger a - \Omega \sigma_z (a\alpha + a^\dagger \alpha^*) - \lambda \sigma_z (a + a^\dagger) + \sum_{j=0}^{n_d} \frac{\epsilon_j}{2} \left[\sigma^+ D(2\alpha) e^{-i\Delta_j t} + \text{H.c.} \right], \tag{A5}$$

where we have neglected a constant energy shift. Therefore, by selecting $\alpha = -\lambda/\Omega$, we obtain a simple Hamiltonian to pursue multiphoton interactions, namely:

$$H_b = \Omega a^\dagger a + \sum_j \frac{\epsilon_j}{2} \left[\sigma^+ e^{2\lambda(a-a^\dagger)/\Omega} e^{-i\Delta_j t} + \text{H.c.} \right], \tag{A6}$$

which is indeed Equation (16). Moving now to an interaction picture w.r.t. $H_{b,0} = (\Omega - \tilde{\nu}) a^\dagger a - \tilde{\omega}\sigma_z/2$, we obtain:

$$H_{b,1}^I = \tilde{\nu} a^\dagger a + \frac{\tilde{\omega}}{2}\sigma_z + \sum_j \frac{\epsilon_j}{2} \left[\sigma^+ e^{-i(\Delta_j + \tilde{\omega})t} e^{2\lambda(a(t) - a^\dagger(t))/\Omega} + \text{H.c.} \right] \tag{A7}$$

with $a(t) = ae^{-i(\Omega-\tilde{\nu})t}$. Requiring $|2\lambda/\Omega|\sqrt{\langle (a+a^\dagger)^2 \rangle} \ll 1$ and selecting $\Delta_j = \Delta_n^{\mp} \equiv \pm n(\tilde{\nu} - \Omega) - \tilde{\omega}$, we resonantly drive multiphoton Jaynes-Cummings interaction terms, while the rest of the terms in the expansion of the exponential term are off-resonant and rotating with a large frequency compared to its amplitude, i.e., $n|\Omega - \tilde{\nu}| \gg \epsilon_j/2$ (for zeroth order) where n is the selected order of the interaction $\sigma^\pm a^n$. In this manner, performing these two approximations, one obtains:

$$H_{\rm n} = \frac{\tilde{\omega}}{2}\sigma_z + \tilde{\nu} a^\dagger a + \sum_{j\in r} \frac{\epsilon_j (2\lambda)^{n_j}}{2\Omega^{n_j} n_j!} \left[\sigma^+ a^{n_j} + \text{H.c.} \right] + \sum_{j \in b} \frac{\epsilon_j (2\lambda)^{n_j}}{2\Omega^{n_j} n_j!} \left[\sigma^+ (-a^\dagger)^{n_j} + \text{H.c.} \right], \tag{A8}$$

where $\Delta_{j\in r} = n_j(\tilde{\nu} - \Omega) - \tilde{\omega}$ and $\Delta_{j \in b} = -n_j(\tilde{\nu} - \Omega) - \tilde{\omega}$, which corresponds to Equation (17). The largest error committed in the previous approximation stems from the zeroth order in the expansion of the exponential. These contributions are of the form $\epsilon_j/2(\sigma^+ e^{in_j(\Omega-\tilde{\nu})t} + \text{H.c.})$, which will produce a significant effect after a time $t \approx n_j(\Omega - \tilde{\nu})/\epsilon_j^2$. For a single n-photon interaction term, population transfer occurs in a characteristic time $\tau_n = \sqrt{n!}(\Omega/2\lambda)^n/\epsilon_0$ (see Section 4.1). Hence, we can provide a rough estimate for the duration of a correct simulation of the desired multiphoton Jaynes-Cummings model to be $t = k\tau_n$ with $k \approx (2\lambda/\Omega)^n n(\Omega - \tilde{\nu})/(\epsilon_0 \sqrt{n!})$.

References

1. Dowling, J.P.; Milburn, G.J. Quantum technology: the second quantum revolution. *Phil. Trans. R. Soc. A* **2003**, *361*, 1655–1674. [CrossRef]
2. Nielsen, M.A.; Chuang, I.L. *Quantum Computation and Quantum Information*; Cambridge University Press: Cambridge, UK, 2000.
3. Feynman, R.P. Simulating physics with computers. *Int. J. Theor. Phys.* **1982**, *21*, 467–488. [CrossRef]
4. Johnson, T.H.; Clark, S.R.; Jaksch, D. What is a quantum simulator? *EPJ Quantum Technol.* **2014**, *1*, 10. [CrossRef]
5. Georgescu, I.M.; Ashhab, S.; Nori, F. Quantum simulation. *Rev. Mod. Phys.* **2014**, *86*, 153–185. [CrossRef]
6. Bloch, I.; Dalibard, J.; Nascimbene, S. Quantum simulations with ultracold quantum gases. *Nat. Phys.* **2012**, *8*, 267–276. [CrossRef]
7. Blatt, R.; Roos, C.F. Quantum simulations with trapped ions. *Nat. Phys.* **2012**, *8*, 277. [CrossRef]
8. Leggett, A.J.; Chakravarty, S.; Dorsey, A.T.; Fisher, M.P.A.; Garg, A.; Zwerger, W. Dynamics of the dissipative two-state system. *Rev. Mod. Phys.* **1987**, *59*, 1–85. [CrossRef]
9. Weiss, U. *Quantum Dissipative Systems*, 3rd ed.; World Scientific: Singapore, 2008.
10. Rabi, I.I. On the process of space quantization. *Phys. Rev.* **1936**, *49*, 324–328. [CrossRef]
11. Rabi, I.I. Space quantization in a gyrating magnetic field. *Phys. Rev.* **1937**, *51*, 652–654. [CrossRef]
12. Jaynes, E.T.; Cummings, F.W. Comparison of quantum and semiclassical radiation theories with application to the beam maser. *Proc. IEEE* **1963**, *51*, 89–109. [CrossRef]
13. Scully, M.O.; Zubairy, M.S. *Quantum Optics*; Cambridge University Press: Cambridge, UK, 1997.
14. Huelga, S.F.; Plenio, M.B. Vibrations, quanta and biology. *Contemp. Phys.* **2013**, *54*, 181–207. [CrossRef]
15. Tanimura, Y.; Kubo, R. Time evolution of a quantum system in contact with a nearly gaussian-Markoffian noise Bath. *J. Phys. Soc. Jpn.* **1989**, *58*, 101–114. [CrossRef]
16. Tanimura, Y. Nonperturbative expansion method for a quantum system coupled to a harmonic-oscillator bath. *Phys. Rev. A* **1990**, *41*, 6676–6687. [CrossRef] [PubMed]
17. Prior, J.; Chin, A.W.; Huelga, S.F.; Plenio, M.B. Efficient simulation of strong system-environment interactions. *Phys. Rev. Lett.* **2010**, *105*, 050404. [CrossRef] [PubMed]
18. Dattani, N.S.; Pollock, F.A.; Wilkins, D.M. Analytic Influence Functionals for Numerical Feynman Integrals in Most Open Quantum Systems. Available online: http://www.naturalspublishing.com/files/published/464k51t1luip94.pdf (accessed on 16 April 2019).
19. Dattani, N.S. FeynDyn: A MATLAB program for fast numerical Feynman integral calculations for open quantum system dynamics on GPUs. *Comput. Phys. Commun.* **2013**, *184*, 2828–2833. [CrossRef]
20. Wilkins, D.M.; Dattani, N.S. Why quantum coherence is not important in the Fenna-Matthews-Olsen complex. *J. Chem. Theor. Comput.* **2015**, *11*, 3411–3419. [CrossRef]
21. Strathearn, A.; Kirton, P.; Kilda, D.; Keeling, J.; Lovett, B.W. Efficient non-Markovian quantum dynamics using time-evolving matrix product operators. *Nat. Commun.* **2018**, *9*, 3322. [CrossRef]
22. Thoss, M.; Wang, H.; Miller, W.H. Self-consistent hybrid approach for complex systems: Application to the spin-boson model with Debye spectral density. *J. Chem. Phys.* **2001**, *115*, 2991–3005. [CrossRef]
23. Martinazzo, R.; Vacchini, B.; Hughes, K.H.; Burghardt, I. Communication: Universal Markovian reduction of Brownian particle dynamics. *J. Chem. Phys.* **2011**, *134*, 011101. [CrossRef]
24. Iles-Smith, J.; Lambert, N.; Nazir, A. Environmental dynamics, correlations, and the emergence of noncanonical equilibrium states in open quantum systems. *Phys. Rev. A* **2014**, *90*, 032114. [CrossRef]
25. Iles-Smith, J.; Dijkstra, A.G.; Lambert, N.; Nazir, A. Energy transfer in structured and unstructured environments: Master equations beyond the Born-Markov approximations. *J. Chem. Phys.* **2016**, *144*, 044110. [CrossRef]
26. Strasberg, P.; Schaller, G.; Lambert, N.; Brandes, T. Nonequilibrium thermodynamics in the strong coupling and non-Markovian regime based on a reaction coordinate mapping. *New J. Phys.* **2016**, *18*, 073007. [CrossRef]
27. Strasberg, P.; Schaller, G.; Schmidt, T.L.; Esposito, M. Fermionic reaction coordinates and their application to an autonomous Maxwell demon in the strong-coupling regime. *Phys. Rev. B* **2018**, *97*, 205405. [CrossRef]

28. Nazir, A.; Schaller, G. The reaction coordinate mapping in quantum thermodynamics. In *Thermodynamics in the Quantum Regime*; Binder, F., Correa, L.A., Gogolin, C., Anders, J., Adesso, G., Eds.; Springer International Publishing: New York, NY, USA, 2018. Available online: https://link.springer.com/book/10.1007/978-3-319-99046-0 (accessed on 16 April 2019).
29. Chin, A.W.; Rivas, Á.; Huelga, S.F.; Plenio, M.B. Exact mapping between system-reservoir quantum models and semi-infinite discrete chains using orthogonal polynomials. *J. Math. Phys.* **2010**, *51*, 092109. [CrossRef]
30. Woods, M.P.; Groux, R.; Chin, A.W.; Huelga, S.F.; Plenio, M.B. Mappings of open quantum systems onto chain representations and Markovian embeddings. *J. Math. Phys.* **2014**, *55*, 032101. [CrossRef]
31. Mascherpa, F.; Smirne, A.; Tamascelli, D.; Fernández-Acebal, P.; Donadi, S.; Huelga, S.F.; Plenio, M.B. Optimized auxiliary oscillators for the simulation of general open quantum systems. *arXiv* **2019**, arXiv:1904.04822.
32. Tamascelli, D.; Smirne, A.; Huelga, S.F.; Plenio, M.B. Nonperturbative treatment of non-Markovian dynamics of open quantum systems. *Phys. Rev. Lett.* **2018**, *120*, 030402. [CrossRef] [PubMed]
33. Braak, D.; Chen, Q.H.; Batchelor, M.T.; Solano, E. Semi-classical and quantum Rabi models: in celebration of 80 years. *J. Phys. A Math. Theor.* **2016**, *49*, 300301. [CrossRef]
34. Lloyd, S.; Braunstein, S.L. Quantum computation over continuous variables. *Phys. Rev. Lett.* **1999**, *82*, 1784–1787. [CrossRef]
35. Felicetti, S.; Pedernales, J.S.; Egusquiza, I.L.; Romero, G.; Lamata, L.; Braak, D.; Solano, E. Spectral collapse via two-phonon interactions in trapped ions. *Phys. Rev. A* **2015**, *92*, 033817. [CrossRef]
36. Pedernales, J.S.; Beau, M.; Pittman, S.M.; Egusquiza, I.L.; Lamata, L.; Solano, E.; del Campo, A. Dirac equation in $(1+1)$-dimensional curved spacetime and the multiphoton quantum Rabi model. *Phys. Rev. Lett.* **2018**, *120*, 160403. [CrossRef]
37. Garbe, L.; Egusquiza, I.L.; Solano, E.; Ciuti, C.; Coudreau, T.; Milman, P.; Felicetti, S. Superradiant phase transition in the ultrastrong-coupling regime of the two-photon Dicke model. *Phys. Rev. A* **2017**, *95*, 053854. [CrossRef]
38. Puebla, R.; Hwang, M.J.; Casanova, J.; Plenio, M.B. Protected ultrastrong coupling regime of the two-photon quantum Rabi model with trapped ions. *Phys. Rev. A* **2017**, *95*, 063844. [CrossRef]
39. Cui, S.; Cao, J.P.; Fan, H.; Amico, L. Exact analysis of the spectral properties of the anisotropic two-bosons Rabi model. *J. Phys. A: Math. Theor.* **2017**, *50*, 204001. [CrossRef]
40. Felicetti, S.; Rossatto, D.Z.; Rico, E.; Solano, E.; Forn-Díaz, P. Two-photon quantum Rabi model with superconducting circuits. *Phys. Rev. A* **2018**, *97*, 013851. [CrossRef]
41. Xie, Y.F.; Duan, L.; Chen, Q.H. Generalized quantum Rabi model with both one- and two-photon terms: A concise analytical study. *Phys. Rev. A* **2019**, *99*, 013809. [CrossRef]
42. Lo, C.F.; Liu, K.L.; Ng, K.M. The multiquantum Jaynes-Cummings model with the counter-rotating terms. *Europhys. Lett.* **1998**, *42*, 1. [CrossRef]
43. Casanova, J.; Puebla, R.; Moya-Cessa, H.; Plenio, M.B. Connecting nth order generalised quantum Rabi models: Emergence of nonlinear spin-boson coupling via spin rotations. *npj Quantum Inf.* **2018**, *4*, 47. [CrossRef]
44. Puebla, R.; Casanova, J.; Houhou, O.; Solano, E.; Paternostro, M. Quantum simulation of multiphoton and nonlinear dissipative spin-boson models. *Phys. Rev. A* **2019**, *99*, 032303. [CrossRef]
45. Breuer, H.P.; Petruccione, F. *The Theory of Open Quantum Systems*; Oxford University Press: Oxford, UK, 2002.
46. Vojta, M. Impurity quantum phase transitions. *Phil. Mag.* **2006**, *86*, 1807–1846. [CrossRef]
47. Hur, K.L. Quantum phase transitions in spin-boson systems: Dissipation and light phenomena. In *Understanding Quantum Phase Transitions*; CRC Press: Boca Raton, FL, USA, 2010; pp. 217–237. Available online: https://www.crcpress.com/Understanding-Quantum-Phase-Transitions/Carr/p/book/9781439802519 (accessed on 16 April 2019).
48. Leibfried, D.; Blatt, R.; Monroe, C.; Wineland, D. Quantum dynamics of single trapped ions. *Rev. Mod. Phys.* **2003**, *75*, 281–324. [CrossRef]
49. Pedernales, J.S.; Lizuain, I.; Felicetti, S.; Romero, G.; Lamata, L.; Solano, E. Quantum Rabi model with trapped ions. *Sci. Rep.* **2015**, *5*, 15472. [CrossRef]

50. Lv, D.; An, S.; Liu, Z.; Zhang, J.N.; Pedernales, J.S.; Lamata, L.; Solano, E.; Kim, K. Quantum simulation of the quantum Rabi model in a trapped Ion. *Phys. Rev. X* **2018**, *8*, 021027. [CrossRef]
51. de Matos Filho, R.L.; Vogel, W. Second-sideband laser cooling and nonclassical motion of trapped ions. *Phys. Rev. A* **1994**, *50*, R1988–R1991. [CrossRef]
52. de Matos Filho, R.L.; Vogel, W. Nonlinear coherent states. *Phys. Rev. A* **1996**, *54*, 4560–4563. [CrossRef]
53. Vogel, W.; de Matos Filho, R.L. Nonlinear Jaynes-Cummings dynamics of a trapped ion. *Phys. Rev. A* **1995**, *52*, 4214–4217. [CrossRef]
54. Cheng, X.H.; Arrazola, I.; Pedernales, J.S.; Lamata, L.; Chen, X.; Solano, E. Nonlinear quantum Rabi model in trapped ions. *Phys. Rev. A* **2018**, *97*, 023624. [CrossRef]
55. de Vega, I.; Alonso, D. Dynamics of non-Markovian open quantum systems. *Rev. Mod. Phys.* **2017**, *89*, 015001. [CrossRef]
56. Breuer, H.P.; Laine, E.M.; Piilo, J. Measure for the degree of non-Markovian behavior of quantum processes in open systems. *Phys. Rev. Lett.* **2009**, *103*, 210401. [CrossRef]

symmetry

MDPI

Article

Parity-Assisted Generation of Nonclassical States of Light in Circuit Quantum Electrodynamics

Francisco A. Cárdenas-López [1,2,*], Guillermo Romero [1], Lucas Lamata [3], Enrique Solano [3,4,5] and Juan Carlos Retamal [1,2,*]

1 Departamento de Física, Universidad de Santiago de Chile (USACH), Avenida Ecuador 3493, Santiago 9170124, Chile; guillermo.romero@usach.cl
2 Center for the Development of Nanoscience and Nanotechnology, Estación Central, Santiago 9170124, Chile
3 Department of Physical Chemistry, University of the Basque Country UPV/EHU, Apartado 644, 48080 Bilbao, Spain; lucas.lamata@gmail.com (L.L.); enr.solano@gmail.com (E.S.)
4 IKERBASQUE, Basque Foundation for Science, Maria Diaz de Haro 3, 48013 Bilbao, Spain
5 Department of Physics, Shanghai University, Shanghai 200444, China
* Correspondence: francisco.cardenas@usach.cl (F.A.C.-L.); juan.retamal@usach.cl (J.C.R.)

Received: 29 January 2019; Accepted: 9 March 2019; Published: 13 March 2019

Abstract: We propose a method to generate nonclassical states of light in multimode microwave cavities. Our approach considers two-photon processes that take place in a system composed of N extended cavities and an ultrastrongly coupled light–matter system. Under specific resonance conditions, our method generates, in a deterministic manner, product states of uncorrelated photon pairs, Bell states, and W states in different modes on the extended cavities. Furthermore, the numerical simulations show that the generation scheme exhibits a collective effect which decreases the generation time in the same proportion as the number of extended cavity increases. Moreover, the entanglement encoded in the photonic states can be transferred towards ancillary two-level systems to generate genuine multipartite entanglement. Finally, we discuss the feasibility of our proposal in circuit quantum electrodynamics. This proposal could be of interest in the context of quantum random number generator, due to the quadratic scaling of the output state.

Keywords: microwave photons; quantum entanglement; superconducting circuits; circuit quantum electrodynamics; quantum Rabi model

1. Introduction

The state-of-the-art devices exhibiting quantum behaviour has grown extensively in the last two decades. Remarkable platforms such as superconducting circuits [1–3] and circuit quantum electrodynamics (QED) [4,5] have allowed the implementation of microwave quantum photonics [6,7], where superconducting electrical circuits mimic the behavior of atoms and cavities [8–10]. In this manner, the capability of tailoring internal circuit parameters to obtain devices with long coherence times and switchable coupling strengths yielded quantum optics experiments such as electromagnetically induced transparency [11], photon blockade [12], and lately to manipulate the parity symmetric of an artificial atom in situ [13] to name a few. A distinctive aspect of microwave photonics is the inherent nonlinearity coming from Josephson junction devices that makes possible to build photonic crystals with Kerr and Cross–Kerr nonlinearities much larger than the one observed in optical devices [14–18]. This allows for enhancing processes such as parametric down conversion [19–22], and the generation of nonclassical states of light [23–27]. Likewise, the notable features of superconducting circuits have also triggered a bunch of proposals for microwave photon generation in systems composed of a large number of cavities. In this context, it is possible to find proposals for the generation of entangled photon states such as NOON (MOON), corresponding to a photonic state where the resonator A has N (M) or zero quanta, entangled with resonator B with zero or N quanta [28–32] states, studies of correlated photons emitted from a cascade system [33], as well as the implementation of a controlled NOT gate (CNOT) gate between qubits encoded in a cavity [34], among other applications [35–37].

On the other hand, circuit QED has also made it possible to achieve light–matter coupling strengths such as the ultrastrong (USC) [38–43] and deep-strong (DSC) [44,45] regimes of light–matter coupling [46,47]. In both cases, as the coupling strength between the light and matter becomes comparable (USC) or larger than the frequency of the field mode (DSC), the rotating wave approximation breaks down and the simplest model that describes the physical situation is the quantum Rabi model [46,48,49]. This model exhibits a discrete parity symmetry and an anharmonic energy spectrum that provide a set of resources for quantum information tasks and quantum simulations [50–60].

Unlike the previous proposal based on microwave photonic state generation, where the considered system works in the single-mode approximation [61–64], and the generation time remains constant independently of the number of subsystems [65], we propose a method to generate nonclassical states of light in multimode microwave cavities. Our approach considers two-photon processes taking place in a system composed of two extended cavities and an ultrastrongly coupled light–matter system, hereafter called quantum Rabi system. Under specific resonance conditions, our method allows a deterministic generation of identical photonic quantum states on different modes, which can be uncorrelated photon state or correlated Bell and W states. Furthermore, we could extend our protocol to N (up to six) cavities. The extension of our system gives rise to a decrease in the generation time of the photonic states. This collective effect arises from the form of the effective coupling obtained in the effective model. In addition, the numerical simulations show that the generation times decrease in the same proportion as the number of extended cavities increases, reducing the detrimental effect due to the interaction of the system with the environment. On the other hand, we show the generation of genuine multipartite entangled states when coupling an ancillary system to each cavity. Finally, we propose a physical implementation of our scheme considering near-term technology of superconducting circuits.

This paper is organized as follows: in Section 2, we introduce our physical scheme. In Section 3, we discuss about the main aspects of the physics of the quantum Rabi system, that is, its parity symmetry and the underlying selection rules for state transitions. In Section 4, we discuss the two-photon processes presented in our physical system, and the generation of nonclassical states of light. In Section 5, we show that our model allows for generating copies of density matrices. In Section 6, we study swapping processes for the generation of genuine multipartite entanglement. In Section 7, we present a physical implementation of our method in superconducting circuits. Finally, in Section 8, we present our concluding remarks.

2. The Model

Let us consider a two-level system of frequency ω_q interacting with a quantized electromagnetic field mode of frequency $\omega_{\rm cav}$ in the USC regime. This system is described by the quantum Rabi Hamiltonian [48,49] ($\hbar = 1$)

$$\mathcal{H}_{\rm QRS} = \omega_{\rm cav} a^{\dagger} a + \frac{\omega_q}{2}\sigma^z + g\sigma^x(a^{\dagger} + a). \tag{1}$$

Here, $a^{\dagger}(a)$ is the creation (annihilation) boson operator for the field mode, the operators σ^x and σ^z are the Pauli matrices describing the two-level system, and g is the light–matter coupling strength. In addition, N two-mode resonators [66], each supporting $M = 2$ modes of frequencies ω_1^{ℓ} and ω_2^{ℓ}, are coupled to the edges of the quantum Rabi system through field quadratures. Notice that each mode couples to the quantum Rabi system with coupling strengths J_1^ℓ and J_2^ℓ, respectively. This physical situation will be described by the Hamiltonian

$$\mathcal{H} = \mathcal{H}_{\rm QRS} + \mathcal{H}_{\rm c} + \mathcal{H}_I, \tag{2}$$

$$\mathcal{H}_{\rm c} = \sum_{\ell=1}^{N} (\omega_1^\ell b_\ell^\dagger b_\ell + \omega_2^\ell c_\ell^\dagger c_\ell), \tag{3}$$

$$\mathcal{H}_I = \sum_{\ell=1}^{N} \left[J_1^\ell (b_\ell^\dagger + b_\ell) + J_2^\ell (c_\ell^\dagger + c_\ell)\right](a + a^\dagger), \tag{4}$$

where $b_\ell^\dagger(b_\ell)$ and $c_\ell^\dagger(c_\ell)$ are the creation (annihilation) boson operators for the first and second field mode of the ℓth cavity, respectively. Notice that the coupling strength between resonators $J_{1,2}^\ell$ can be several orders of magnitude smaller than $\omega_{1,2}^\ell$ [67]. Hence, the counter-rotating terms present in Equation (4) can be neglected through the rotating wave approximation (RWA) [68] leading to the following interaction Hamiltonian:

$$\mathcal{H}_I = \sum_{\ell=1}^{N} \left[(J_1^\ell b_\ell + J_2^\ell c_\ell) a^\dagger + (J_1^\ell b_\ell^\dagger + J_2^\ell c_\ell^\dagger) a\right]. \tag{5}$$

In what follows, we will discuss the features of the energy spectrum of the quantum Rabi system, that is, its anharmonicity and the internal symmetry arising in the USC regime.

3. Parity Symmetry $\mathbb{Z}_2$ and Selection Rules

The energy spectrum of the quantum Rabi system presents interesting features, which promises to be useful for quantum information processing [50–55]. These features correspond to the anharmonicity of the energy levels and the selection rules imposed by the $\mathbb{Z}_2$ symmetry arising in the USC regime. In Figure 1, we show the first four energy levels of quantum Rabi system as a function of $g/\omega_{\rm cav}$, where we see an anharmonic energy spectrum. Moreover, in the quantum Rabi system, it is possible to define the parity operator $\mathcal{P} = -\sigma^z \otimes e^{i\pi a^\dagger a}$ that has a discrete spectrum $p = \pm 1$. Notice that $\mathcal{P}$ commutes with the quantum Rabi system Hamiltonian, $[H_{\rm QRS}, \mathcal{P}] = 0$, thus enabling the diagonalization of both operators in a common basis $\{|E, p\rangle\}_{E=0}^{\infty}$. We label each quantum state regarding two quantum numbers, E corresponds to the energy level while p denotes its parity value. In Figure 1, states with parity $+1(-1)$ are denoted by the continuous orange (dashed blue) line. As a consequence, the Hilbert space of the quantum Rabi system is divided into two parts, the even and the odd parity subspaces. This allows, depending on the kind of driving, the possibility of connecting states with different or equal parity. For instance, it has been proven that drivings like $\mathcal{H}_D \sim (a^\dagger + a)$ and $\mathcal{H}_D \sim \sigma^x$ connect states belonging to different subspaces [55]. This happens because the matrix element $\langle E, \pm|\mathcal{H}_D|E', \mp\rangle \neq 0$. Moreover, for a driving like $\mathcal{H}_D \sim \sigma^z$, only states with equal parity can be connected since the matrix element $\langle E, \pm|\mathcal{H}_D|E', \pm\rangle \neq 0$.

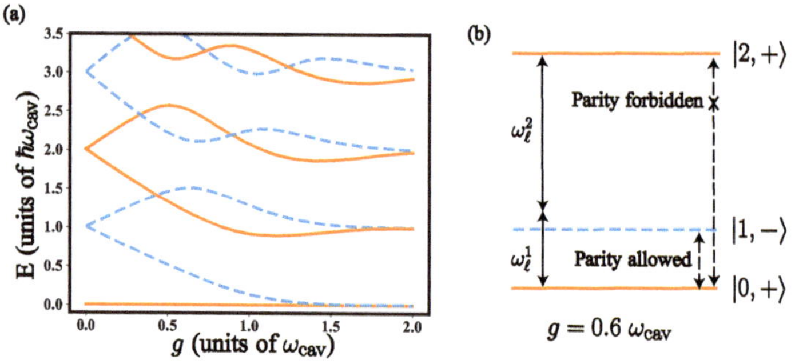

Figure 1. (**a**) energy spectrum of the Hamiltonian in Equation (1) as a function of the coupling strength g. Blue dashed lines stand for states with parity $p = +1$. Orange continuous lines correspond to states with parity $p = -1$; (**b**) diagram of the energy levels at $g = 0.6\ \omega_{\text{cav}}$. In these numerical calculations, we use $\omega_q = 0.8\ \omega_{\text{cav}}$.

4. Two Photon Process Mediated by a Quantum Rabi System

Here, we propose the implementation of a two-photon process mediated by the quantum Rabi system, which relies on its anharmonicity and the selection rules previously discussed. In particular, we provide specific resonance conditions between the two-mode cavities and the quantum Rabi system to achieve the phase matching condition analogue to the usual parametric down-conversion process in optical systems.

Let us consider the following set of parameters for quantum Rabi system $\omega_q = 0.8\ \omega_{\text{cav}}$ and $g = 0.6\ \omega_{\text{cav}}$. In this case, as shown in Figure 1, the first three energy levels form a cascade Ξ system similar to Rydberg atoms studied in cavity quantum electrodynamics [69,70]. The ground and second excited state have parity $p = +1$, while the first excited state has parity $p = -1$ (see Figure 1b). Notice that this behaviour on the energy levels is valid for $g < 0.4\ \omega_{\text{cav}}$. Otherwise, the parity value of the lowest energy levels does not resemble a cascade energy configuration. In such a case, it is not possible to implement a two-photon process. According to the type of interaction of the two-mode cavities with the quantum Rabi system, see Equation (4), a single photon will not be able to produce a transition between the second excited state $|2,+\rangle$ and the ground state $|0,+\rangle$ since it is forbidden by parity. However, these states can be connected through a second-order process. The latter may occur when the sum of frequencies of the modes, belonging to a cavity, matches that of the energy transition between the ground and the second excited state of the quantum Rabi system, i.e., $\omega_1^\ell + \omega_2^\ell = \nu_{20}$. Moreover, the frequency of each mode must be far-off-resonance with respect to the frequency of the first excited state $\omega_{1,2}^\ell \gg \nu_{10}$. Under these conditions, the intermediate level can be adiabatically eliminated leading to the effective Hamiltonian

$$\mathcal{H}_{\text{eff}}^\ell = \mathcal{H}_{\text{QRS}} + \mathcal{H}_{\text{c}} + \sum_{\ell,\ell'=1}^{N} \mathcal{J}_\ell^{\ell'} (b_\ell^\dagger c_{\ell'}^\dagger \mathcal{S}^- + b_\ell c_{\ell'} \mathcal{S}^+), \tag{6}$$

which describes simultaneous two-photon processes in both cavities. Here, $S^+ = |2,+\rangle\langle 0,+|$ corresponds to the ladder operator of the quantum Rabi system in the effective two-level basis. Furthermore, the effective coupling strength $\mathcal{J}_\ell^{\ell'}$ is defined as follows:

$$\mathcal{J}_\ell^{\ell'} = J_1^\ell J_2^{\ell'} \chi_{01}\chi_{21} \left[\frac{1}{\Delta_{10}^1} + \frac{1}{\Delta_{21}^2} \right]. \tag{7}$$

Here, we define the matrix element of the operator a in the quantum Rabi system basis as $\chi_{kj}^{\pm} = \langle k,+|a|j,-\rangle$ and the quantum Rabi system-mode detuning $\Delta_{kj}^{1,2} = \omega_{1,2}^\ell - \nu_{kj}$. The Hamiltonian in Equation (6) gives rise to several parametric down conversion processes mediated by the quantum Rabi system, i.e., by starting with one excitation on the quantum Rabi system of energy ν_{20}, it may produce a pair of photons of frequencies ω_1 and ω_2. The photons generated by this scheme will distribute on the two-mode cavities according to the relation $\omega_1^{\ell} + \omega_2^{\ell'} = \nu_{20}$. Depending on the number of cavities N, this condition enables us to generate two uncorrelated single-photons ($N = 1$), or produce identical entangled states of different frequencies such as Bell states ($N = 2$) or W states ($N \geq 3$). For the cases, $N = \{1, 2, 3\}$, the effective Hamiltonians read

$$\mathcal{H}_{\rm eff}^{1} = \mathcal{J}_2^1 \left[b_1^\dagger c_1^\dagger \mathcal{S}^- + b_1 c_1 \mathcal{S}^+\right], \tag{8a}$$

$$\mathcal{H}_{\rm eff}^{2} = \mathcal{J}_2^1 \left[b_1^\dagger c_1^\dagger + b_2^\dagger c_2^\dagger + b_1^\dagger c_2^\dagger + b_2^\dagger c_1^\dagger\right] \mathcal{S}^- + \text{H.c}, \tag{8b}$$

$$\mathcal{H}_{\rm eff}^{3} = \mathcal{J}_2^1 \left[b_1^\dagger c_1^\dagger + b_2^\dagger c_2^\dagger + b_3^\dagger c_3^\dagger + b_1^\dagger c_2^\dagger + b_1^\dagger c_3^\dagger + b_2^\dagger c_1^\dagger + b_2^\dagger c_3^\dagger + b_3^\dagger c_1^\dagger + b_3^\dagger c_2^\dagger\right] \mathcal{S}^- + \text{H.c.} \tag{8c}$$

The protocol works as follows: we initially consider the entire system in its ground state i.e., $|\Psi(0)\rangle = |0, +\rangle \bigotimes_{\ell,\ell'}^{N} |0_\ell, 0_{\ell'}\rangle$. Afterwards, one may excite the quantum Rabi system with a microwave pulse with frequency $\nu = \nu_{20}$. Notice that ν_{20} is not resonant with the frequency of the two-mode resonators. Thus, the resonator modes coupled dispersively to the quantum Rabi system remaining in the vacuum state. This interaction can be modelled by the Hamiltonian $\mathcal{H}_D = \Omega\cos(\nu_{20}t)\sigma^z$. Notice that $\mathcal{H}_D$ preserves the $\mathbb{Z}_2$ symmetry of the quantum Rabi system, thus enabling transitions between states of equal parity. The state of the system, after an interaction time $t = \pi/\Omega$, is given by $|\Psi(\pi/\Omega)\rangle = |2, +\rangle \bigotimes_{\ell,\ell'}^{N} |0_\ell, 0_{\ell'}\rangle$. Then, the system evolves under the Hamiltonian (2) for a time $t_S = \pi/(2\mathcal{J}_2^1)$, $t_B = \pi/(4\mathcal{J}_2^1)$, or $t_W = \pi/(6\mathcal{J}_2^1)$, for generating uncorrelated single photons, pair of Bell states, or pair of W states, respectively. As a result, the quantum Rabi system excitation generates two photons distributed on the cavities satisfying the relation $\omega_1^{\ell} + \omega_2^{\ell'} = \nu_{20}$. The wave functions of the system after algebraic manipulation read

$$|\Psi(\pi/\Omega + \pi/2\mathcal{J}_2^1)\rangle_S = |+, 0\rangle \otimes |1_{\omega_1}\rangle \otimes |1_{\omega_2}\rangle, \tag{9a}$$

$$|\Psi(\pi/\Omega + \pi/4\mathcal{J}_2^1)\rangle_B = |+, 0\rangle \otimes |\Psi_{\omega_1}^+\rangle \otimes |\Psi_{\omega_2}^+\rangle, \tag{9b}$$

$$|\Psi(\pi/\Omega + \pi/6\mathcal{J}_2^1)\rangle_W = |+, 0\rangle \otimes |W_{\omega_1}\rangle \otimes |W_{\omega_2}\rangle, \tag{9c}$$

where $|\Psi_{\omega_n}^+\rangle$ is the Bell state for photons of frequency ω_n distributed over different resonators, that is, $|\Psi_{\omega_n}^+\rangle = \frac{1}{\sqrt{2}}[|1_{\omega_n}\rangle|0_{\omega_n}\rangle + |0_{\omega_n}\rangle|1_{\omega_n}\rangle]$. In addition, the state $|W_{\omega_n}\rangle$ stands for a W state of a single photon of frequency ω_n distributed over different cavities. For the case of three cavities, $|W_{\omega_n}\rangle$ is given by [71]

$$|W_{\omega_n}\rangle = \frac{1}{\sqrt{3}}\left(|1_{\omega_n}\rangle|0_{\omega_n}\rangle|0_{\omega_n}\rangle + |0_{\omega_n}\rangle|1_{\omega_n}\rangle|0_{\omega_n}\rangle + |0_{\omega_n}\rangle|0_{\omega_n}\rangle|1_{\omega_n}\rangle\right). \tag{10}$$

This state represents one photon of frequency ω_n which can be distributed over three different cavities In Figure 2, we show the numerical calculations of the above-mentioned protocol. Here, we compute the population evolution of states $|\Psi(0)\rangle$, and states $|\Psi\rangle_S$, $|\Psi\rangle_B$, and $|\Psi\rangle_W$ given in Equation (9). The parametric interaction can produce either uncorrelated photon states of different frequency or identical entangled states of modes belonging to distinct cavities. Furthermore, the simulations show that the state generation time decreases as $1/N$. This can be explained by analysing the structure of Equation (8). As the effective Hamiltonians describe a quantum dynamics in a reduced two-dimensional Hilbert space, the matrix elements between the initial state $|\Psi(0)\rangle$ and $|\Psi\rangle_S$, $|\Psi\rangle_B$, and $|\Psi\rangle_W$ are proportional to the normalization of the desired state, that is, $\sqrt{N} \times \sqrt{N}$, where $N = 1$ stands for single photons, $N = 2$ for Bell states, and $N \geq 3$ for W states. In other words, the matrix elements of the effective Hamiltonians are proportional to the number of two-mode cavities. By considering the following parameters for the quantum Rabi system, $\omega_{\rm cav} = 2\pi \times 13.12$ GHz [38], qubit frequency $\omega_q = 0.8\omega_{\rm cav}$, and light–matter coupling strength $g = 0.6\omega_{\rm cav}$, we can estimate $|\chi_{10}| = 0.8188$ and $|\chi_{21}| = 1.235$. In addition, we choose $\omega_1^n = 0.25\nu_{20}$,

$\omega_2^n = 0.75\nu_{20}$, $J_1^n = 0.0075\nu_{20}$, and $J_2^n = 0.0053\nu_{20}$. In this case, the state generation times are about $t_S \approx 25.10(8)$ (ns), $t_B \approx 12.55(4)$ (ns), $t_W \approx 8.369(4)$ (ns) for $N = 3$, and $t_W \approx 6.28$ (ns) for $N = 4$ (see Figure 2).

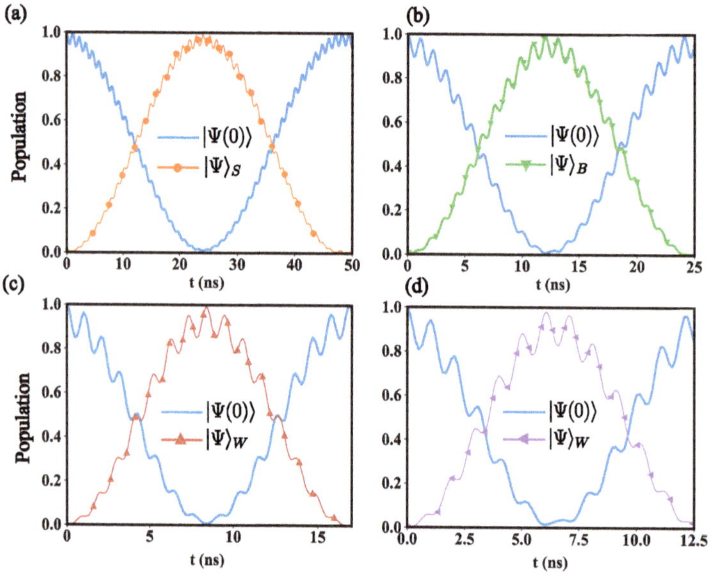

Figure 2. Population evolution of the Hamiltonian in Equation (2) for initial state $|\Psi(0)\rangle = |2,+\rangle \otimes_{\ell,n}^{N,M} |0_\ell^n\rangle$ with cases $N = 1$ (**a**), $N = 2$ (**b**), $N = 3$ (**c**), and $N = 4$ (**d**) two-mode cavities. Blue continuous line is the evolution of the initial state $|\Psi(0)\rangle$. (**a**) orange dotted line denotes the population of $|\Psi\rangle_S = |0,+\rangle \otimes |1_{\omega_1}\rangle \otimes |1_{\omega_2}\rangle$; (**b**) green dotted line stands for the population of $|\Psi\rangle_B = |0,+\rangle \otimes |\Psi_{\omega_1}^+\rangle \otimes |\Psi_{\omega_2}^+\rangle$; and (**c**) red dotted line stands for $|\Psi\rangle_W = |0,+\rangle \otimes |W_{\omega_1}\rangle \otimes |W_{\omega_2}\rangle$; (**d**) purple dotted line stand for the $|\Psi\rangle_W = |0,+\rangle \otimes |W_{\omega_1}\rangle \otimes |W_{\omega_2}\rangle$, where this W contains four modes. The parameters for these calculations can be found in the main text.

It is interesting to mention that Bell and GHZ states have been proposed to be generated in coupled systems in the USC regime of cQED [72]. The authors consider the two-level system and the field modes as separate entities. In such a case, the USC regime only contributes to counter-rotating terms allowing multi-photon interaction terms. Our work considers the USC system formed by a field mode and a qubit as a whole. Thus, the properties on the energy spectrum of the USC system allow us to generate multi-photon states by coupling the USC system to two-mode resonators in a second order process to specific resonance conditions. Finally, our scheme allows for generating copies of W states spatially distributed in the two-mode resonator setups.

5. Copies of Density Matrices

In the above section, we have demonstrated that our system can generate identical copies of pure microwave photon states ($N = 1, 2, 3$). Here, we demonstrate that even including loss mechanisms our protocol can still generate copies of density matrices with high fidelity. Since our proposal includes an ultrastrongly coupled light–matter system, the dissipative dynamics will be described by the master equation [73]

$$\begin{aligned}\dot{\rho}(t) &= i[\rho(t), \mathcal{H}] + \sum_{\ell=1}^{N} \kappa_\ell \mathcal{D}[b_\ell]\rho(t) + \sum_{\ell=1}^{N} \kappa_\ell \mathcal{D}[c_\ell]\rho(t) \\ &+ \sum_{E,E>E'} (\Gamma_\kappa^{EE'} + \Gamma_\gamma^{EE'} + \Gamma_{\gamma_\phi}^{EE'})\mathcal{D}[|E,p\rangle\langle E',p'|]\rho(t).\end{aligned} \tag{11}$$

Here, $\mathcal{H}$ is the Hamiltonian of Equation (2) and $\mathcal{D}[O]\rho = 1/2(2O\rho O^\dagger - \rho O^\dagger O - O^\dagger O \rho)$ is the Liouvillian operator. Furthermore, κ_ℓ^n stands for photon loss rate for each cavity mode. $\Gamma_\kappa^{EE'}$, $\Gamma_\gamma^{EE'}$ and $\Gamma_{\gamma_\phi}^{EE'}$ are the dressed decay rates associated with the quantum Rabi system, and they are defined as $\Gamma_\kappa^{EE'} = \frac{\kappa}{\omega_{\rm cav}} \nu_{EE'} |X_{EE'}|^2$, $\Gamma_\gamma^{EE'} = \frac{\gamma}{\omega_q} \nu_{EE'} |\sigma_{EE'}^x|^2$ and $\Gamma_{\gamma_\phi}^{EE'} = \frac{\gamma_\phi}{\omega_q} \nu_{EE'} |\sigma_{EE'}^z|^2$, where κ, γ and γ_ϕ are the bare photon leakage, relaxation, and depolarizing noise rates, respectively. In the derivation of the master equation, it has to be assumed that the spectral densities describing the system–environment interactions correspond to an ohmic bath [74,75]. In this case, the impedance $Z(\omega)$ of each circuit element can be modelled as a resistor [76].

To study the robustness of our protocol under loss mechanisms, first we will examine the generation of copies of density matrices for the cases of $N = 1, 2, 3$ two-mode cavities. As mentioned in the previous section, the whole system is initialized in the state $|\Psi(0)\rangle = |0,+\rangle \otimes_{\ell,\ell'}^N |0_\ell, 0_{\ell'}\rangle$. Then, we let the system to evolve under Equation (11) for three different times: $t_S = \pi/(2\mathcal{J}_2^1)$, $t_B = \pi/(4\mathcal{J}_2^1)$, and $t_W = \pi/(6\mathcal{J}_2^1)$, for $N = 1$, $N = 2$, and $N = 3$ two-mode cavities, respectively. Once the corresponding density matrix $\rho(t)$ is obtained, we trace over the quantum Rabi system and modes ω_2 (ω_1) to obtain the reduced density matrix ρ_{ω_1} (ρ_{ω_2}), which contains only degrees of freedom associated with the mode ω_1 (ω_2) distributed on different two-mode cavities. Table 1, first row, shows the fidelity between both reduced density matrices $\mathcal{F}(\rho_{\omega_1}, \rho_{\omega_2}) = \mathrm{Tr}(\rho_{\omega_1}\rho_{\omega_2})$. These results allow us to conclude that both quantum states are identical up to 99% fidelity for a single cavity, and up to 98% fidelity for two and three cavities. Table 1 also shows the fidelities of generating the states of Equation (9), which is, $\mathcal{F}_S = \mathrm{Tr}(\rho(t_S)\rho_S)$, $\mathcal{F}_B = \mathrm{Tr}(\rho(t_B)\rho_S)$, and $\mathcal{F}_W = \mathrm{Tr}(\rho(t_W)\rho_S)$, where $\rho(t)$ have been numerically calculated from Equation (11). The high fidelities of our protocol are mainly due to the fast state generation times as compared with the loss rates. Our numerical calculations have been carried out with realistic circuit QED parameters at temperature T = 15 mK [77]. For the quantum Rabi system decay rates, we consider values $\kappa = 2\pi \times 0.10$ MHz, $\gamma = 2\pi \times 15$ MHz and $\gamma_\phi = 2\pi \times 7.69$ MHz and for the cavities $\kappa_\ell^n = \kappa$.

The way to cease the system dynamics once we have obtained the entangled states is to tune the frequency of the two-level system forming the QRS. In such a case, the QRS becomes far off-resonant with the two-mode cavities, and the state does not evolve anymore. The time at which the system maintains the quantum state must be of the order of the decay time of the cavity. We do not expect that the decay time of the QRS affects this process, due to the fact that the QRS is in its ground state $|0,+\rangle$.

Table 1. Summarized fidelity values between the states ρ_{ω_ℓ} obtained through of the master Equation (17) with the fictitious states $\rho_{\rm probe}$ and $\rho_{\rm tensor}$ for the case where the quantum Rabi system is coupled to $n = \{1, 2, 3\}$ two-mode cavity.

	$N=1$	$N=2$	$N=3$
$\mathcal{F}(\rho_{\omega_1}, \rho_{\omega_2})$	0.9898	0.9818	0.9832
$\mathcal{F}_S$	0.9892	-	-
$\mathcal{F}_B$	-	0.9945	-
$\mathcal{F}_W$	-	-	0.9904

6. Entanglement Swapping between Distant Superconducting Qubits

In this section, we study the transfer of entanglement generated into the field modes towards distant superconducting circuits. Let us consider a pair of two-level systems coupled at the end of each cavity. As we shall see later in Section 7, our physical implementation will consider $\lambda/4$ transmission line resonators, and superconducting flux qubits to guarantee strong coupling between them. In such a case, we describe the system with the following Hamiltonian

$$\mathcal{H}_{\text{ES}} = \mathcal{H} + \sum_{\ell=1}^{2} \frac{\omega_{q\ell}^{n}}{2}\sigma_{\ell}^{z} + \sum_{\ell=1}^{2} \lambda_{\ell}\sigma_{\ell}^{x}(b_{\ell} + b_{\ell}^{\dagger}) + \lambda_{\ell}^{'}\sigma_{\ell}^{x}(c_{\ell} + c_{\ell}^{\dagger}), \tag{12}$$

where $\mathcal{H}$ is the Hamiltonian defined in Equation (2). Moreover, σ_{ℓ}^{x} and σ_{ℓ}^{z} are Pauli matrices describing the two-level systems, $\{b_{\ell}, c_{\ell}\}$, are the annihilation boson operators of the extended cavities. Additionally, λ_{ℓ}, and $\lambda_{\ell}^{'}$ are the coupling strength between the two-level system with the first and second field mode cavity, respectively. Depending on whether the two-level systems are resonant with either mode ω_{ℓ}^{1} or ω_{ℓ}^{2}, the process with the coupling strength λ_{ℓ} or $\lambda_{\ell}^{'}$ becomes dispersive $|\omega_{\ell}^{1} - \omega_{q\ell}^{n}| \gg \{\lambda_{\ell}, \lambda_{\ell}^{'}\}$ [78], and therefore we neglect it via the rotating wave approximation. The following master equation describes the system dynamics

$$\dot{\rho}(t) = i[\rho(t), \mathcal{H}_{\text{ES}}] + \sum_{\ell=1}^{N} \gamma_{\ell}\mathcal{D}[\sigma_{\ell}^{-}]\rho(t) + \sum_{\ell=1}^{N} \gamma_{\phi_{\ell}}\mathcal{D}[\sigma_{\ell}^{z}]\rho(t). \tag{13}$$

The last two terms describe the loss mechanisms acting on the two-level system, i.e., relaxation on the qubit at a rate γ and depolarizing noise at rate γ_{ϕ}. The entanglement swapping protocol is the following; we initialize the whole system in its ground state

$$\rho_0 = |0,+\rangle\langle 0,+| \bigotimes_{\ell,\ell'}^{N} |0_{\ell}, 0_{\ell'}\rangle\langle 0_{\ell}, 0_{\ell'}| \bigotimes_{\ell}^{N} |g_{\ell}\rangle\langle g_{\ell}|. \tag{14}$$

We dispersively couple the two-level systems with the field modes on the cavities ($|\omega_{\ell}^{1,2} - \omega_{q\ell}| \gg (\lambda_{\ell}, \lambda_{\ell}^{'})$) [78]. Then, we drive the quantum Rabi system to prepare it in the second excited state $|2,+\rangle$

$$\rho_1 = |2,+\rangle\langle 2,+| \bigotimes_{\ell,\ell'}^{N} |0_{\ell}, 0_{\ell'}\rangle\langle 0_{\ell}, 0_{\ell'}| \bigotimes_{\ell}^{N} |g_{\ell}\rangle\langle g_{\ell}|. \tag{15}$$

This state is the initial condition of our scheme. Afterwards, we let the system evolve under the Hamiltonian in Equation (13). Due to the dispersive qubit–resonator interaction, the two-level systems do not evolve. After a time $t = \pi/(2\mathcal{J}_{\text{eff}})$, the density matrix of the system reads

$$\rho_2 = |0,+\rangle\langle 0,+| \bigotimes_{\ell}^{N} |\Psi_{\omega_{\ell}}^{+}\rangle\langle \Psi_{\omega_{\ell}}^{+}| \bigotimes_{\ell}^{N} |g_{\ell}\rangle\langle g_{\ell}|. \tag{16}$$

The next step is to avoid the generated photons coming back to the quantum Rabi system. To achieve it, we tune far-off resonance the quantum Rabi system and the resonators by changing the qubit frequency that belongs to the quantum Rabi system. Afterwards, we put into resonance the external two-level system with either ω_{ℓ}^{1} or ω_{ℓ}^{2} field modes. In such a case, for a time $t = \pi/(2\lambda_{\ell})$ ($t = \pi/(2\lambda_{\ell}^{'})$), the system evolves to

$$\rho_3 = |0,+\rangle\langle 0,+| \bigotimes_{\ell=1}^{N} |0_{\omega_1^\ell}\rangle\langle 0_{\omega_1^\ell}| \otimes |\Psi_{\omega_2^\ell}\rangle\langle\Psi_{\omega_2^\ell}| \otimes |\Phi\rangle\langle\Phi|, \tag{17}$$

$$\rho_3 = |0,+\rangle\langle 0,+| \bigotimes_{\ell=1}^{N} |\Psi_{\omega_2^\ell}\rangle\langle\Psi_{\omega_2^\ell}| \otimes |0_{\omega_2^\ell}\rangle\langle 0_{\omega_2^\ell}| \otimes |\Phi\rangle\langle\Phi|. \tag{18}$$

Here, $|\Phi\rangle = (|g_1e_2\rangle + |e_1g_2\rangle)/\sqrt{2}$ is a Bell state of the pair of qubits. This protocol is illustrated in Figure 3. On the other hand, Figure 4 shows the real and imaginary part of the reduced density matrix for the pair of qubits after performing the protocol. As the figure shows, even though the loss mechanisms act on the system, the entanglement of the modes can be transferred to the qubits with high fidelity. For the two-level systems coupled to the first mode (ω_1^ℓ), the fidelity is $\mathcal{F} = 0.9960$, and $\mathcal{F} = 0.9976$ when the qubit is resonant with the second mode (ω_2^ℓ). This transfer occurs at the time scale of $t_{S_1} = 23.08$ [ns] and $t_{S_2} = 16.32$ [ns], respectively.

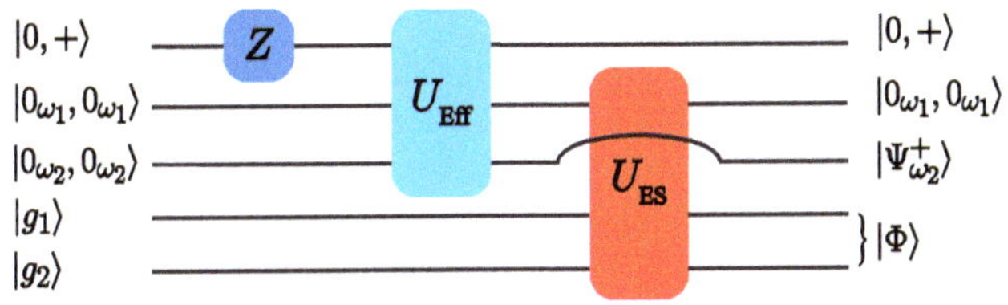

Figure 3. Gate sequence for the entanglement swapping protocol. At first, the quantum Rabi system is initialized from $|0,+\rangle$ to $|2,+\rangle$ via a driving acting on σ^z. Afterwards, the system evolves under the gate $U_{\rm eff} = \exp(-it\mathcal{H}_{\rm eff}/\hbar)$. Then, the auxiliary two-level systems are tuned to the mode ω_1 (ω_2). Thus, the system starts to evolve under $\mathcal{H}_{ES}$ to entangle the qubits.

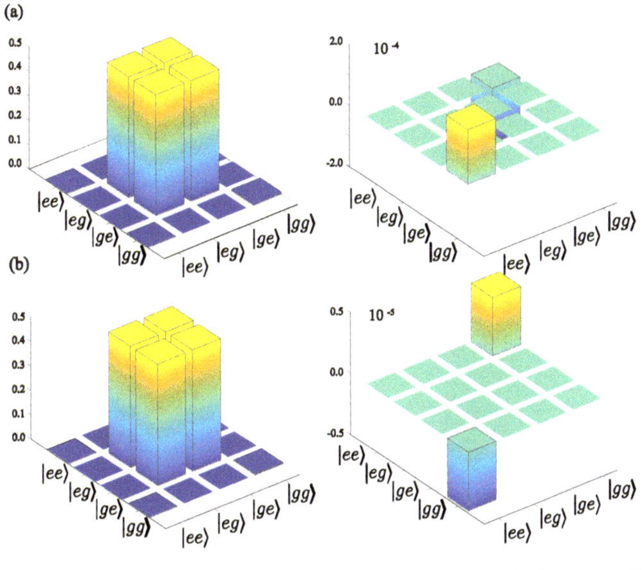

Figure 4. Real and imaginary part of the reduced density matrix composed of the two qubits coupled to the field mode of frequency ω_1 (**a**) and mode ω_2 (**b**). The fidelity between the simulated state and the Bell state $|\Phi\rangle = (|eg\rangle + |eg\rangle)/\sqrt{2}$ is (**a**) $\mathcal{F} = 0.9960$ and (**b**) $\mathcal{F} = 0.9976$.

7. Implementation in Circuit QED

We depict the schematic implementation of our system in Figure 5. The circuit is composed of a non-uniform $\lambda/2$ transmission line resonator of length d galvanically coupled to a four-junction flux qubit at the middle of the resonator. The non-uniform shape of the resonator produces an increasing on its inductance in the vicinity of the qubit. In addition, the additional junction on the flux-qubit in the shared wire also produces an increase on the inductance of the resonator. As a result, the qubit-resonator coupling strength can achieve the USC regime [38]. Moreover, at the edges of this $\lambda/2$ resonator, one may couple up to six additional $\lambda/4$ transmission line resonator also of length d via capacitances. The capacitive coupling follows the same procedure as in Ref. [79]. In such a case, the finger pattern between the superconducting metal and the substrate form the capacitive coupling at the end of these resonators. The orthogonal arrangement between the two-mode cavity reduces the crosstalk between these resonators, reducing the cavity–cavity interaction. The Lagrangian representing this situation for $N = 2$ resonators (extension to more two-mode resonators is straightforward) reads

$$\mathcal{L} = \mathcal{L}_{\text{QRS}} + \mathcal{L}_{\text{c}} + \mathcal{L}_{I}, \tag{19}$$

where $\mathcal{L}_{\text{QRS}}$ is the quantum Rabi system Lagrangian constituted by the $\lambda/2$ transmission line resonator coupled to a four-junction flux qubit, $\mathcal{L}_{\text{c}}$ is the two-mode $\lambda/4$ transmission line resonator Lagrangian, whereas $\mathcal{L}_{I}$ stands for the resonator-resonator coupling Lagrangian obtained from the capacitive coupling. The quantum Rabi system Lagrangian is given by

$$\mathcal{L}_{\text{QRS}} = \int_0^d dz\left[\frac{c}{2}[\partial_t\psi(z,t)]^2 - \frac{1}{2l}[\partial_z\psi(z,t)]^2\right] + \sum_{k=1}^{4}\left[\frac{C_{J,k}}{2}\dot{\varphi}_k^2 + E_{J,k}\cos\left(\frac{\varphi_k}{\phi_0}\right)\right]. \tag{20}$$

Here, $\psi(z,t)$, and φ_k correspond to the flux nodes for the $\lambda/2$ transmission line resonator and the four-Josephson flux qubit, respectively. These variables are related with the voltage drop through the specific branch component by means the relation $\psi(z,t) = \int_{-\infty}^{t} V(t')dt'$ [80,81]. Furthermore, c and l are the capacitance and inductance per unit length of the resonator, while $C_{J,k}$ and $E_{J,k}$ are the capacitance and energy describing the k-th Josephson junction. The two-mode resonator Lagrangian is given by

$$\mathcal{L}_{\text{c}} = \sum_{\ell=1}^{N=2}\left\{\int_0^d dz\left[\frac{c_\ell}{2}[\partial_t\phi_\ell(z,t)]^2 - \frac{1}{2l_\ell}[\partial_z\phi_\ell(z,t)]^2\right]\right\} + \frac{C_r}{2}[\partial_t\phi_1(d,t)]^2 + \frac{C_r}{2}[\partial_t\phi_2(0,t)]^2, \tag{21}$$

where ϕ_ℓ, $\ell = 1,2$ is the flux node describing the ℓth two-mode resonator. Moreover, c_ℓ, and l_ℓ stand for the capacitance and inductance per unit length of the ℓth two-mode resonator. Furthermore, C_r is the coupling capacitance between the two-mode resonators with the QRS resonator. Finally, $\mathcal{L}_I$ is the interaction Lagrangian given by

$$\mathcal{L}_I = -C_r\left[\dot{\phi}_1(d,t)\dot{\psi}(0,t) + \dot{\psi}(d,t)\dot{\phi}_2(0,t)\right]. \tag{22}$$

As the two-level system with the resonator forming the quantum Rabi system is ultrastrongly coupled, we will expect in principle that the qubit also couples with the two-mode resonators. However, this does not occur due to the nature of the coupling between the flux-qubit with the transmission line resonator; as the $\lambda/2$ couples to the flux-qubit through the current, the latter should be placed at the position where the current reaches its maximum to achieve the USC coupling regime. In the $\lambda/2$ resonator, this position corresponds to the centre of the line. Thus, the edges of the QRS resonator have zero current and the qubit two-mode resonator coupling vanishes. As a result, the two-mode resonators couple to the QRS only through the resonator.

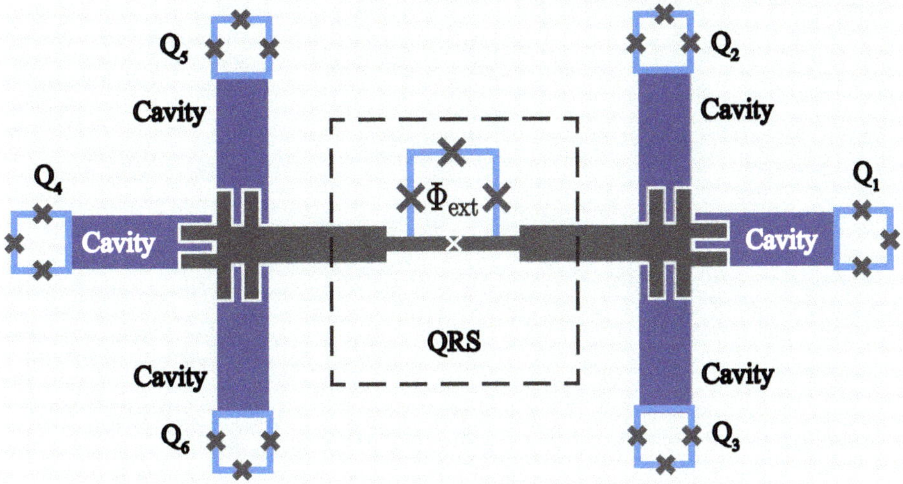

Figure 5. Schematic illustration of our superconducting circuit implementation. Here, the quantum Rabi system is composed of a $\lambda/2$ transmission line resonator (grey resonator) interacting with a superconducting flux qubit located at the middle point to achieve the USC regime. In addition, the $\lambda/2$ resonator is coupled at its edges forming a finger pattern to two-mode transmission lines (blue resonators) through capacitive coupling. The limitation to keep up to six resonators relies on the reduction of the crosstalk between the resonators. The crosstalk induces a mutual-inductance effect that leads to a resonator–resonator coupling given by the following Hamiltonian. Furthermore, at the end of the two-mode transmission line resonator superconducting flux qubit Q_ℓ are coupled.

7.1. Rabi System Hamiltonian

For this derivation, we assume $E_{J,1} = E_{J,2} = E_J$, $E_{J,3} = \alpha E_J$ and $E_{J,4} = \gamma E_J$. Moreover, the fluxoid quantization relation on the superconducting loop is given by

$$\varphi_1 - \varphi_2 + \varphi_3 + \varphi_4 = -2\pi f_x, \tag{23}$$

where f_x is the frustration parameter defined as $f_x = \phi_{\text{ext}}/\Phi_0$. On the other hand, we assume that the Josephson inductance of the fourth junction is smaller than the inductance of the flux-qubit loop, thus most of the current flowing through the resonator [38]. As a consequence, the qubit acts as a small perturbation of the transmission line resonator. Thus, the phase difference is given by $\varphi_4 = \Delta\psi$, where $\Delta\psi = \psi(z_i, t) - \psi(z_{i-1}, t)$ corresponds to the phase difference of the $\lambda/2$ transmission line resonator at the position where the four Josephson junction is placed. Thus, the Lagrangian takes the following form

$$\begin{aligned}\mathcal{L}_{\text{QRS}} &= \int_0^d dz\left[\frac{c}{2}[\partial_t\psi(z,t)]^2 - \frac{1}{2l}[\partial_z\psi(z,t)]^2\right] + \frac{C_J}{2}\left[\dot{\varphi}_1{}^2 + \dot{\varphi}_2^2 + \alpha(\dot{\varphi}_2 - \dot{\varphi}_1 - \Delta\dot{\psi})^2 + \gamma\Delta\dot{\psi}^2\right] \\ &+ E_J\left[\cos\left(\frac{\varphi_1}{\phi_0}\right) + \cos\left(\frac{\varphi_2}{\phi_0}\right) + \gamma\cos\left(\frac{\Delta\psi}{\phi_0}\right) + \alpha\cos\left(\frac{\varphi_2 - \varphi_1 + \phi_{\text{ext}} - \Delta\psi}{\phi_0}\right)\right].\end{aligned} \tag{24}$$

We are assuming the superconducting phase on the loop is well localized, thus the potential energy can be expanded in powers of $\Delta\psi/\phi_0$ [51], allowing us to express the quantum Rabi system Lagrangian in the following form

$$\mathcal{L}_{\text{QRS}} = \mathcal{L}_{\text{r}} + \mathcal{L}_{\text{q}} + \mathcal{L}_{\text{qr}}, \tag{25}$$

where $\mathcal{L}_{\rm r}$ is the Lagrangian of the resonator with an embedded junction

$$\mathcal{L}_{\rm r} = \int_0^d dz \left[\frac{c}{2}[\partial_t \psi(z,t)]^2 - \frac{1}{2l}[\partial_z \psi(z,t)]^2\right] + \frac{C_J(\alpha+\gamma)}{2}\Delta\dot{\psi} + \gamma E_J \cos\left(\frac{\Delta\psi}{\phi_0}\right). \tag{26}$$

Moreover, $\mathcal{L}_{\rm q}$ is the usual three-junction flux qubit Lagrangian [8]

$$\mathcal{L}_{\rm q} = \frac{C_J}{2}\left[(1+\alpha)(\dot{\varphi_1}^2 + \dot{\varphi}_2^2) - 2\alpha\dot{\varphi}_2\dot{\varphi}_1\right] + E_J\left[\cos\left(\frac{\varphi_1}{\phi_0}\right) + \cos\left(\frac{\varphi_2}{\phi_0}\right) + \alpha\cos\left(\frac{\varphi_2 - \varphi_1 + \phi_{\rm ext}}{\phi_0}\right)\right]. \tag{27}$$

Finally, $\mathcal{L}_{\rm qr}$ is the qubit-resonator Lagrangian; this term has two contributions: capacitive and galvanic coupling, and reads

$$\mathcal{L}_{\rm qr} = -\alpha C_J(\dot{\varphi_1} + \dot{\varphi_2})\Delta\dot{\psi} - \frac{\alpha E_J}{\phi_0}\sin\left(\frac{\varphi_1 - \varphi_2 + \phi_{\rm ext}}{\phi_0}\right)\Delta\psi. \tag{28}$$

In the flux qubit, the capacitive energy is smaller than the inductive energy [38]. Thus, we neglect the capacitive term, obtaining

$$\mathcal{L}_{\rm qr} = -\frac{\alpha E_J}{\phi_0}\sin\left(\frac{\varphi_1 - \varphi_2 + \phi_x}{\phi_0}\right)\Delta\psi. \tag{29}$$

We obtain the Lagrangian for the transmission line resonator by computing its equation of motion. In such a case, the flux $\psi(z,t)$ obeys the wave equation whose solution for the $\lambda/2$ transmission line resonator is given by

$$\psi(z,t) = \sum_m \mathcal{U}_m(z)\mathcal{G}_m(t), \tag{30}$$

$$\psi(z,t) = \sum_m \left[A_m \cos k_m(z - d/2) + B_m \sin k_m(z + d/2)\right]\mathcal{G}_m(t), \tag{31}$$

where k_m is the wave vector of the resonator with the embedded junction, which is obtained through the dispersion relation

$$k_m \tan\left(\frac{k_m d}{2}\right) = \frac{2l}{L_J}\left[1 - \left(\frac{vk_m}{\omega_p}\right)^2\right], \tag{32}$$

with $v = \sqrt{1/lc}$ is the transmission line resonator wave velocity, $L_J = \gamma\phi_0^2/E_J$ is the Josephson inductance. In addition, $\omega_p = 1/\sqrt{L_J C_J}$ is the plasma frequency of the embedded junction. Replacing the flux $\psi(z,t)$ on the Lagrangian given in Equation (21), we arrive at

$$\mathcal{L}_{\rm r} = \sum_m \left[\frac{\eta_m \dot{\mathcal{G}}_m(t)^2}{2} - \frac{\eta_m^2\omega_m^2\mathcal{G}_m^2(t)}{2}\right], \tag{33}$$

where η_m is the effective capacitance [17]. By applying the Legendre transformation, we arrive at the classical Hamiltonian

$$\mathcal{H}_{\rm r} = \sum_m \left[\frac{\Pi_m^2}{2\eta_m} + \frac{\eta_m^2\omega_m^2 G_m^2}{2}\right]. \tag{34}$$

Here, $\Pi_m = \partial\mathcal{L}/\partial[\dot{G}_m]$ is the canonical conjugate momenta. We proceed to quantize the Hamiltonian promoting the following operators:

$$\Pi_m = \sqrt{\frac{\hbar}{2\eta_m\omega_m}}(a_m^\dagger + a_m), \tag{35}$$

$$G_m = i\sqrt{\frac{\hbar\eta_m\omega_m}{2}}(a_m^\dagger - a_m). \tag{36}$$

Replacing these operators in the Hamiltonian $\mathcal{H}_r$ we arrive at the transmission line resonator quantum Hamiltonian

$$\mathcal{H}_\mathrm{r} = \sum_m \hbar\omega_m\left(a_m^\dagger a_m + \frac{1}{2}\right). \tag{37}$$

Now, let us consider the Lagrangian of the four-junction flux qubit given in Equation (27). Close to the degeneracy point $\phi_x = \phi_0/2$, the system can be truncated to the two lowest eigenstates, whose Hamiltonian is given by

$$\mathcal{H}_\mathrm{q} = \frac{\hbar\omega_\mathrm{q}}{2}\sigma^z, \tag{38}$$

where $\omega_\mathrm{q} = \sqrt{\Delta^2 + \varepsilon^2}$, with Δ the qubit gap, and $\varepsilon = 2I_p(\phi_x - \phi_0/2)$, where I_p is the persistent current on the superconducting loop. Furthermore, the interacting Lagrangian given in Equation (29) can be written in the two-level basis, in such case, the quantized Hamiltonian reads

$$\mathcal{H}_\mathrm{qr} = i\frac{\alpha E_J \Delta\mathcal{U}_m}{\phi_0}\sqrt{\frac{\hbar\eta_m\omega_m}{2}}S_{01}\sigma^x(a_m - a_m^\dagger), \tag{39}$$

$$S_{01} = \langle 0|\left[\sin\left(\frac{\varphi_1 - \varphi_2 + \phi_x}{\phi_0}\right)\right]|1\rangle. \tag{40}$$

Thus, the quantum Rabi Hamiltonian is given by

$$\mathcal{H}_\mathrm{QRS} = \sum_m \hbar\omega_m a_m^\dagger a_m + \frac{\hbar\omega_q}{2}\sigma^z + \hbar\sum_m g_m\sigma^x(a_m^\dagger + a_m). \tag{41}$$

Notice that the coupling strength between the transmission line resonator and the artificial atom depends on two factors: the position at which the two-level system is placed, and the nature of the coupling, i.e., galvanic or capacitive. In our case, as the artificial atom corresponds to a flux-qubit, it is coupled to the current on the transmission line resonator. As a consequence, the two-level system only couples to even modes because the odd modes have a node [39] in the flux qubit position as illustrated in Figure 6a. The spectrum of the multi-mode Rabi system is depicted in Figure 6b. Notice that the energy spectrum of the multimode quantum Rabi system preserves the parity symmetry exhibited by the single mode quantum Rabi system (see Figure 1). Furthermore, for a wide range of coupling strength g, the low-lying energy states exhibit the same selection rules observed in the single mode Rabi system. Thus, by adding more complexity to the mediator system (quantum Rabi system), our proposed generation scheme is still useful due to the preserve of the selection rules on the system:

$$\mathcal{H}_\mathrm{QRS} = \hbar\omega_\mathrm{cav}a^\dagger a + \frac{\hbar\omega_q}{2}\sigma^z + \hbar g\sigma^x(a^\dagger + a). \tag{42}$$

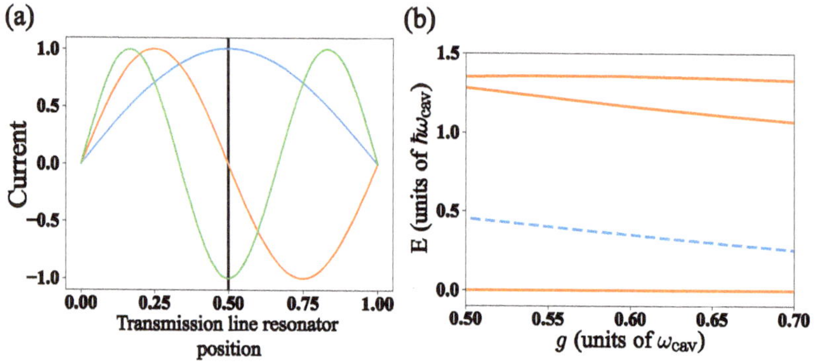

Figure 6. (**a**) sketch of the current distribution of the first three resonator modes for the $\lambda/2$ transmission line resonator. The vertical black line corresponds to the position at which the artificial atom is placed. (**b**) Energy spectrum of the Hamiltonian in Equation (41) considering the first three field modes. Orange lines corresponds to energy levels with parity $p = +1$, whereas blue dashed line stands for energy levels with parity $p = -1$.

7.2. Multimode Cavity Hamiltonian

To obtain the Hamiltonian of the two-mode cavities, let us consider the Lagrangian given in Equation (21) for $N = 2$ resonators

$$\mathcal{L}_c = \sum_{\ell=1}^{N=2}\left\{\int_0^d dz\left[\frac{c_\ell}{2}[\partial_t\phi_\ell(z,t)]^2 - \frac{1}{2l_\ell}[\partial_z\phi_\ell(z,t)]^2\right]\right\} + \frac{C_r}{2}[\partial_t\phi_1(d,t)]^2 + \frac{C_r}{2}[\partial_t\phi_2(0,t)]^2. \quad (43)$$

For the specific implementation, we consider boundary conditions defining a $\lambda/4$ resonator, where the current at the ends where the two-mode resonator coupled to the QRS resonator is zero, and the voltage reaches its maximum. These conditions are given by

$$-\partial_z\phi_1(0,t) = -\partial_z\phi_2(d,t) = 0, \quad (44)$$

$$\partial_t\phi_1(d,t) = \partial_t\phi_2(0,t) = 0. \quad (45)$$

By solving the wave equation with the previous boundary conditions, we obtain the expression for the flux on the ℓth $\lambda/4$ transmission line resonator

$$\phi_1 = \sum_n A_n\cos(q_{n,\ell}z)G_n(t); \qquad \phi_2 = \sum_n B_n\cos q_{n,\ell}(z-d)G_n(t), \quad (46)$$

where G_n satisfy the time-dependent part of the wave equation. Moreover, $q_{n,\ell}$ corresponds to a quasi-momentum satisfying the following dispersion relation:

$$q_{n,\ell} = \frac{1}{v_\ell l_\ell C_r}\cot(q_n d). \quad (47)$$

By replacing the expression of the fluxes ϕ_1, and ϕ_2 and performing the Legendre transformation, we arrive at the quantum $\lambda/4$ transmission line resonator Hamiltonian

$$\mathcal{H}_c = \sum_{\ell=1}^{N}\sum_n \hbar\omega^{\ell,n}\left(b_{\ell,n}^\dagger b_{\ell,n} + \frac{1}{2}\right). \quad (48)$$

Here, the index n runs over all the mode on the transmission line, whereas the index ℓ stands for the number of multi-mode resonator coupled to the QRS. Notice that in principle all the modes of the $\lambda/4$ are involved in the system dynamics. However, numerical simulations show that, due to the resonance condition on our system, the field mode greater than three does not induce dynamics in the system. Thus, by keeping the notation given in Equation (1), we rewrite the Hamiltonian in Equation (48) as follows:

$$\mathcal{H}_c = \sum_{\ell=1}^{N} \left[\hbar\omega_1^\ell \left(b_\ell^\dagger b_\ell + \frac{1}{2} \right) + \hbar\omega_2^\ell \left(c_\ell^\dagger c_\ell + \frac{1}{2} \right) \right], \tag{49}$$

where ω_1^ℓ and ω_2^ℓ correspond to the frequency of the first and second field mode of the ℓth $\lambda/4$, respectively. Likewise, b_ℓ and c_ℓ are the boson operator for the first and second field mode, respectively.

7.3. Complete Model

With the Hamiltonian of the free system already obtained in the previous section, we are able to write the interacting Hamiltonian of the complete system. Before the application of the Legendre transformation, the Lagrangian in Equation (22) turns on the interaction Hamiltonian

$$\mathcal{H}_I = C_r \left[\dot{\phi}_1(d,t)\dot{\psi}(0,t) + \dot{\psi}(d,t)\dot{\phi}_2(0,t) \right]. \tag{50}$$

By replacing the expression of the fluxes $\psi(z,t)$ and $\phi_\ell(z,t)$ already obtained, we arrive at the quantum interaction Hamiltonian

$$\mathcal{H}_I = \sum_{\ell=1}^{N} \sum_{m} [J_1^{\ell,m}(b_\ell^\dagger + b_\ell) + J_2^{\ell,m}(c_\ell^\dagger + c_\ell)](a_m + a_m^\dagger), \tag{51}$$

$$J_1^{\ell,m} = \frac{\hbar C_r}{2} \sqrt{\frac{1}{\eta_m \bar{\eta}_{n,\ell} \omega_m \omega_1^\ell}} \psi(0)\phi_1(d), \tag{52}$$

$$J_2^{\ell,m} = \frac{\hbar C_r}{2} \sqrt{\frac{1}{\eta_m \bar{\eta}_{n,\ell} \omega_m \omega_2^\ell}} \psi(d)\phi_1(0). \tag{53}$$

Here, $\bar{\eta}_{n,\ell}$ is the effective capacitance of the $\lambda/4$ transmission line resonator. Notice that capacitive coupling strength between resonators is commonly at least one order of magnitude smaller than the bare frequency of the field mode frequency. Thus, for resonator in the Giga Hertz regime, coupling strength $J_{1,2}^{\ell,m}$ are within the Mega Hertz regime. In the single mode approximation for the QRS resonator, we obtain

$$\mathcal{H}_I = \sum_{\ell=1}^{N} [J_1^\ell(b_\ell^\dagger + b_\ell) + J_2^\ell(c_\ell^\dagger + c_\ell)](a + a^\dagger). \tag{54}$$

Thus, the complete system Hamiltonian is given by

$$\begin{aligned} \mathcal{H} = {} & \hbar\omega_{\text{cav}} a^\dagger a + \frac{\hbar\omega_q}{2}\sigma^z + \hbar g \sigma^x (a^\dagger + a) + \sum_{\ell=1}^{N} \left[\hbar\omega_1^\ell \left(b_\ell^\dagger b_\ell + \frac{1}{2} \right) + \hbar\omega_2^\ell \left(c_\ell^\dagger c_\ell + \frac{1}{2} \right) \right] \\ & + \sum_{\ell=1}^{N} [J_1^\ell(b_\ell^\dagger + b_\ell) + J_2^\ell(c_\ell^\dagger + c_\ell)](a + a^\dagger). \end{aligned} \tag{55}$$

To assure that our approximations are valid, we compute the system dynamics for the case of $N = 1$ resonator that contains three modes. As we see in Figure 7, due to the third mode, it is not resonant with the QRS energy transition. This contribution does not affect the generation scheme.

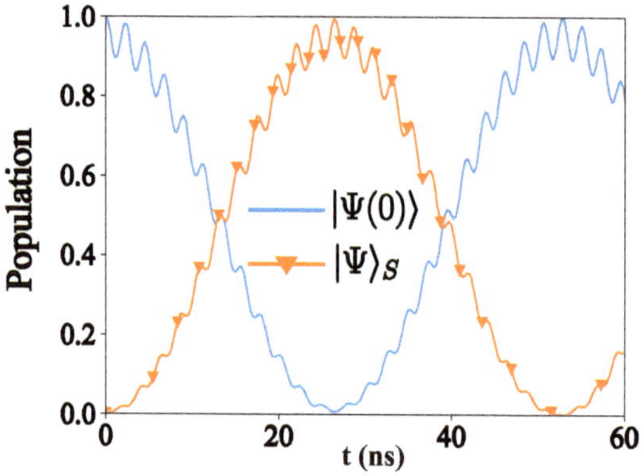

Figure 7. Population evolution of the Hamiltonian in Equation (2) for the case where the multi-mode resonator contains three modes. The system is prepared in the state $|\Psi(0)\rangle = |2,+\rangle \otimes_{\ell,n}^{N,M} |0_\ell^n\rangle$. The blue continuous line is the evolution of the initial state $|\Psi(0)\rangle$. The orange dotted line denotes the population of $|\Psi\rangle_S = |0,+\rangle \otimes |1_{\omega_1}\rangle \otimes |1_{\omega_2}\rangle$. The parameters for these calculations can be found in the main text.

7.4. Driving the Superconducting Qubit

We can drive the two-level system by applying a time-dependent magnetic field on the superconducting loop (see Figure 5). In such case, the energy gap ω_q can be expressed as

$$\omega_q(t) = \sqrt{\Delta^2 + \varepsilon^2(t)}, \tag{56}$$

where $\varepsilon(t) = \varepsilon_{\rm DC} + \varepsilon_{\rm AC}\cos(\omega_L t)$ is the time-dependent energy on the system, which contains DC and AC contributions [82]. For $\varepsilon_{DC} \gg \varepsilon_{AC}$, we can write the flux-qubit energy as

$$\omega_q = \sqrt{\Delta^2 + \varepsilon_{DC}^2} + \frac{\varepsilon_{\rm DC}\varepsilon_{\rm AC}}{\sqrt{\Delta^2 + \varepsilon_{DC}^2}}\cos(\omega_L t). \tag{57}$$

Thus, the flux-qubit driving Hamiltonian is given by

$$\mathcal{H}_q(t) = \frac{\omega_q}{2}\sigma^z + \Omega\cos(\omega_L t)\,\sigma^z. \tag{58}$$

8. Conclusions

In summary, we have shown the usefulness of the quantum Rabi system to generate photons under suitable configuration. Based on the selection rules and the anharmonicity present in the quantum Rabi system, it is possible to find the specific matching condition for producing two-photon processes, analogous to the observed in the parametric frequency conversion. This condition allows us to generate in a deterministic manner uncorrelated or correlated photon states, Bell and W states. The protocol mentioned above, together with available optical to microwave photon converter technologies, may be a useful resource to perform tasks as distributed quantum computing or quantum cryptography.

On the other hand, the proposed protocol could work as a quantum random number generator (QRNG) in the microwave regime. Unlike the optical regime where QRNGs are based on single mode and polarization states of photons, our proposal considers two-mode states of photons. As a consequence, we observe a quadratic increase in the amount of possible quantum random numbers that would be generated in comparison with the single-mode case. Moreover, due to the fact that our system generates simultaneously identical maximally entangled photonic states of different frequency, this state resembles a N^2-side dice, where each side is associated with the probability to find the photons of frequency ω_1 and ω_2 in one of the two modes in N cavities. Thus, the multiphoton process mediated by the quantum Rabi system occurring on the two-mode cavities provides an efficient way to produce quantum random numbers. This efficiency relies on two main aspects of our protocol. The former is concerned with the collective effect producing a decrease of the generation time as the number of cavities increases, permitting the avoidance of the bias produced by the interaction of the system with the environment. The latter concerns the multimode configuration of our scheme. As we previously mentioned, the inclusion of the multimode systems allows us to increase the amount of possible quantum random numbers as the number of devices required decrease. Finally, we have also proposed a possible experimental implementation of our scheme considering near-term technology on circuit quantum electrodynamics in the ultrastrong coupling regime.

Author Contributions: F.A.C.-L. performed all the calculations. All authors contributed to the generation of ideas, development of the research, and writing of the manuscript.

Funding: The authors acknowledge support from CEDENNA, Financiamiento Basal para Centros Científicos y Tecnológicos de Excelencia FB.0807, Dirección de Postgrado USACH, FONDECYT Grant No. 1150653 and No. 1140194, Spanish MINECO/FEDER FIS2015-69983-P, Basque Government IT986-16, and Ramón y Cajal Grant RYC-2012-11391. This material is also based upon work supported by the projects OpenSuperQ and QMiCS of the EU Flagship on Quantum Technologies, and by the U.S. Department of Energy, Office of Science, Office of Advanced Scientific Computing Research (ASCR) quantum algorithm teams program, under field work proposal number ERKJ333.

Acknowledgments: We thank Leong-Chuan Kwek for fruitful discussions.

Conflicts of Interest: The authors declare no conflict of interest.

References

1. You, J.Q.; Nori, F. Atomic physics and quantum optics using superconducting circuits. *Nature* **2011**, *474*, 589–597. [CrossRef] [PubMed]
2. Houck, A.A.; Türeci, H.E.; Koch, J. On-chip quantum simulation with superconducting circuits. *Nat. Phys.* **2012**, *8*, 292–299. [CrossRef]
3. Devoret, M.H.; Schoelkopf, R.J. Superconducting circuits for quantum information: An outlook. *Science* **2013**, *339*, 1169–1174. [CrossRef]
4. Wallraff, A.; Schuster, D.I.; Blais, A.; Frunzio, L.; Huang, R.S.; Majer, J.; Kumar, S.; Girvin, S.M.; Schoelkopf, R.J. Strong coupling of a single photon to a superconducting qubit using circuit quantum electrodynamics. *Nature* **2004**, *431*, 162–167. [CrossRef] [PubMed]
5. Blais, A.; Huang, R.S.; Wallraff, A.; Girvin, S.M.; Schoelkopf, R.J. Cavity quantum electrodynamics for superconducting electrical circuits: An architecture for quantum computation. *Phys. Rev. A* **2004**, *69*, 062320. [CrossRef]
6. Nakamura, Y. Microwave quantum photonics in superconducting circuits. In Proceedings of the IEEE Photonics Conference 2012, Burlingame, CA, USA, 23–27 September 2012; pp. 544–545.
7. Gu, X.; Kockum, A.F.; Miranowicz, A.; Liu, Y.X.; Nori, F. Microwave photonics with superconducting quantum circuits. *Phys. Rep.* **2017**, *718–719*, 1–102. [CrossRef]
8. Orlando, T.P.; Mooij, J.E.; Tian, L.; van der Wal, C.H.; Levitov, L.S.; Lloyd, S.; Mazo, J.J. Superconducting persistent-current qubit. *Phys. Rev. B* **1999**, *60*, 15398–15413. [CrossRef]
9. Koch, J.; Terri, M.Y.; Gambetta, J.; Houck, A.A.; Schuster, D.I.; Majer, J.; Blais, A.; Devoret, M.H.; Girvin, S.M.; Schoelkopf, R.J. Charge-insensitive qubit design derived from the Cooper pair box. *Phys. Rev. A* **2007**, *76*, 042319. [CrossRef]
10. Göppl, M.; Fragner, A.; Baur, M.; Bianchetti, R.; Filipp, S.; Leek, J.M.F.P.J.; Puebla, G.; Steffen, L.; Wallraff, A. Coplanar waveguide resonators for circuit quantum electrodynamics. *J. Appl. Phys.* **2008**, *104*, 113904. [CrossRef]
11. Abdumalikov, A.A.; Astafiev, O.; Zagoskin, A.M.; Pashkin, Y.A.; Nakamura, Y.; Tsai, J.S. Electromagnetically induced transparency on a single artificial atom. *Phys. Rev. Lett.* **2010**, *104*, 193601. [CrossRef]
12. Lang, C.; Bozyigit, D.; Eichler, C.; Steffen, L.; Fink, J.M.; Abdumalikov, A.A., Jr.; Baur, M.; Filipp, S.; Da Silva, M.P.; Blais, A.; et al. Observation of resonant photon blockade at microwave frequencies using correlation function measurements. *Phys. Rev. Lett.* **2011**, *106*, 243601. [CrossRef]

13. Goetz, J.; Deppe, F.; Fedorov, K.G.; Eder, P.; Fischer, M.; Pogorzalek, S.; Xie, E.; Marx, A.; Gross, R. Parity-Engineered Light-Matter Interaction. *Phys. Rev. Lett.* **2018**, *121*, 060503. [CrossRef]
14. Bergeal, N.; Vijay, R.; Manucharyan, V.E.; Siddiqi, I.; Schoelkopf, R.J.; Girvin, S.M.; Devoret, M.H. Analog information processing at the quantum limit with a Josephson ring modulator. *Nat. Phys.* **2010**, *6*, 296–302. [CrossRef]
15. Boissonneault, M.; Gambetta, J.M.; Blais, A. Improved superconducting qubit readout by qubit-induced nonlinearities. *Phys. Rev. Lett.* **2010**, *105*, 100504. [CrossRef] [PubMed]
16. Bourassa, J.; Beaudoin, F.; Gambetta, J.M.; Blais, A. Josephson-junction-embedded transmission-line resonators: From Kerr medium to in-line transmon. *Phys. Rev. A* **2012**, *86*, 013814. [CrossRef]
17. Leib, M.; Deppe, F.; Marx, A.; Gross, R.; Hartmann, M.J. Networks of nonlinear superconducting transmission line resonators. *New J. Phys.* **2012**, *14*, 075024. [CrossRef]
18. Hoi, I.-C.; Kockum, A.F.; Palomaki, T.; Stace, T.M.; Fan, B.; Tornberg, L.; Sathyamoorthy, S.R.; Johansson, G.; Delsing, P.; Wilson, C.M. Giant Cross–Kerr Effect for Propagating Microwaves Induced by an Artificial Atom. *Phys. Rev. Lett.* **2013**, *111*, 053601. [CrossRef] [PubMed]
19. Marquardt, F. Efficient on-chip source of microwave photon pairs in superconducting circuit QED. *Phys. Rev. B* **2007**, *76*, 205416. [CrossRef]
20. Koshino, K. Down-conversion of a single photon with unit efficiency. *Phys. Rev. A* **2009**, *79*, 013804. [CrossRef]
21. Liu, Y.X.; Sun, H.C.; Peng, Z.H.; Miranowicz, A.; Tsai, J.S.; Nori, F. Controllable microwave three-wave mixing via a single three-level superconducting quantum circuit. *Sci. Rep.* **2014**, *4*, 7289. [CrossRef]
22. Sánchez-Burillo, E.; Martín-Moreno, L.; García-Ripoll, J.J.; Zueco, D. Full two-photon down-conversion of a single photon. *Phys. Rev. A* **2016**, *94*, 053814. [CrossRef]
23. Yurke, B.; Corruccini, L.R.; Kaminsky, P.G.; Rupp, L.W.; Smith, A.D.; Silver, A.H.; Simon, R.W.; Whittaker, E.A. Observation of parametric amplification and deamplification in a Josephson parametric amplifier. *Phys. Rev. A* **1989**, *39*, 2519–2533. [CrossRef]
24. Everitt, M.J.; Clark, T.D.; Stiffell, P.B.; Vourdas, A.; Ralph, J.F.; Prance, R.J.; Prance, H. Superconducting analogs of quantum optical phenomena: Macroscopic quantum superpositions and squeezing in a superconducting quantum-interference device ring. *Phys. Rev. A* **2004**, *69*, 043804. [CrossRef]
25. Zagoskin, A.M.; Il'chev, E.; McCutcheon, M.W.; Young, J.F.; Nori, F. Controlled generation of squeezed states of microwave radiation in a superconducting resonant circuit. *Phys. Rev. Lett.* **2008**, *101*, 253602. [CrossRef]
26. Moon, K.; Girvin, S.M. Theory of microwave parametric down-conversion and squeezing using circuit QED. *Phys. Rev. Lett.* **2005**, *95*, 140504. [CrossRef]
27. Didier, N.; Qassemi, F.; Blais, A. Perfect squeezing by damping modulation in circuit quantum electrodynamics. *Phys. Rev. A* **2014**, *89*, 013820. [CrossRef]
28. Strauch, F.W. All-resonant control of superconducting resonators. *Phys. Rev. Lett.* **2012**, *109*, 210501. [CrossRef] [PubMed]
29. Zhao, Y.-J.; Wang, C.; Zhu, X.; Liu, Y.-X. Engineering entangled microwave photon states through multiphoton interactions between two cavity fields and a superconducting qubit. *Sci. Rep.* **2016**, *6*, 23646. [CrossRef]
30. Merkel, S.T.; Wilhelm, F.K. Generation and detection of NOON states in superconducting circuits. *New J. Phys.* **2010**, *12*, 093036. [CrossRef]
31. Strauch, F.W.; Jacobs, K.; Simmonds, R.W. Arbitrary Control of Entanglement between two Superconducting Resonators. *Phys. Rev. Lett.* **2010**, *105*, 050501. [CrossRef] [PubMed]
32. Wang, H.; Mariantoni, M.; Bialczak, R.C.; Lenander, M.; Lucero, E.; Neeley, M.; O'Connell, A.D.; Sank, D.; Weides, M.; Wenner, J.; et al. Deterministic Entanglement of Photons in Two Superconducting Microwave Resonators. *Phys. Rev. Lett.* **2011**, *106*, 060401. [CrossRef]
33. Gasparinetti, S.; Pechal, M.; Besse, J.C.; Mondal, M.; Eichler, C.; Wallraff, A. Correlations and entanglement of microwave photons emitted in a cascade decay. *Phys. Rev. Lett.* **2017**, *119*, 140504. [CrossRef] [PubMed]
34. Campagne-Ibarcq, P.; Zalys-Geller, E.; Narla, A.; Shankar, S.; Reinhold, P.; Burkhart, L.D.; Axline, C.J.; Pfaff, W.; Frunzio, L.; Schoelkopf, R.J.; et al. Deterministic remote entanglement of superconducting circuits through microwave two-photon transitions. *Phys. Rev. Lett.* **2018**, *120*, 200501. [CrossRef] [PubMed]
35. Rosenblum, S.; Gao, Y.Y.; Reinhold, P.; Wang, C.; Axline, C.J.; Frunzio, L.; Girvin, S.M.; Jiang, L.; Mirrahimi, M.; Devoret, M.H.; et al. A CNOT gate between multiphoton qubits encoded in two cavities. *Nat. Commun.* **2018**, *9*, 652. [CrossRef] [PubMed]

36. Narla, A.; Shankar, S.; Hatridge, M.; Leghtas, Z.; Sliwa, K.M.; Zalys-Geller, E.; Mundhada, S.O.; Pfaff, W.; Frunzio, L.; Shoelkopf, R.J.; et al. Robust concurrent remote entanglement between two superconducting qubits. *Phys. Rev. X* **2016**, *6*, 031036. [CrossRef]
37. Kurpiers, P.; Magnard, P.; Walter, T.; Royer, B.; Pechal, M.; Heinsoo, J.; Salathé, Y.; Akin, A.; Storz, S.; Besse, J.C.; et al. Deterministic Quantum State Transfer and Generation of Remote Entanglement using Microwave Photons. *Nature* **2018**, *558*, 264–267. [CrossRef]
38. Bourassa, J.; Gambetta, J.M.; Abdumalikov, A.A.; Astafiev, O.; Nakamura, Y.; Blais, A. Ultrastrong coupling regime of cavity QED with phase-biased flux qubits. *Phys. Rev. A* **2009**, *80*, 032109. [CrossRef]
39. Niemczyk, T.; Deppe, F.; Huebl, H.; Menzel, E.P.; Hocke, F.; Schwarz, M.J.; García-Ripoll, J.J.; Zueco, D.; Hümmer, T.; Solano, E.; et al. Circuit quantum electrodynamics in the ultrastrong-coupling regime. *Nat. Phys.* **2010**, *6*, 772–776. [CrossRef]
40. Forn-Díaz, P.; Lisenfeld, J.; Marcos, D.; García-Ripoll, J.J.; Solano, E.; Harmans, C.J.P.M.; Mooij, J.E. Observation of the Bloch-Siegert shift in a qubit-oscillator system in the ultrastrong coupling regime. *Phys. Rev. Lett.* **2010**, *105*, 237001. [CrossRef]
41. Andersen, C.K.; Blais, A. Ultrastrong coupling dynamics with a transmon qubit. *New J. Phys.* **2017**, *19*, 023022. [CrossRef]
42. Forn-Díaz, P.; García-Ripoll, J.J.; Peropadre, B.; Orgiazzi, J.L.; Yurtalan, M.A.; Belyansky, R.; Wilson, C.M.; Lupascu, A. Ultrastrong coupling of a single artificial atom to an electromagnetic continuum in the nonperturbative regime. *Nat. Phys.* **2017**, *13*, 39–43. [CrossRef]
43. Martínez, J.P.; Léger, S.; Gheeraert, N.; Dassonneville, R.; Planat, L.; Foroughi, F.; Krupko, Y.; Buisson, O.; Naud, C.; Hasch-Guichard, W.; et al. A tunable Josephson platform to explore many-body quantum optics in circuit-QED. *arXiv* **2018**, arXiv:1802.00633.
44. Casanova, J.; Romero, G.; Lizuain, I.; García-Ripoll, J.J.; Solano, E. Deep Strong Coupling Regime of the Jaynes-Cummings Model. *Phys. Rev. Lett.* **2010**, *105*, 263603. [CrossRef]
45. Yoshihara, F.; Fuse, T.; Ashhab, S.; Kakuyanagi, K.; Saito, S.; Semba, K. Superconducting qubit-oscillator circuit beyond the ultrastrong-coupling regime. *Nat. Phys.* **2017**, *13*, 44–47. [CrossRef]
46. Forn-Díaz, P.; Lamata, L.; Rico, E.; Kono, J.; Solano, E. Ultrastrong coupling regimes of light–matter interaction. *arXiv* **2018**, arXiv:1804.09275.
47. Kockum, A.F.; Miranowicz, A.; de Liberato, S.; Savasta, S.; Nori, F. Ultrastrong coupling between light and matter. *Nat. Rev. Phys.* **2019**, *1*, 19–40. [CrossRef]
48. Rabi, I.I. On the Process of Space Quantization. *Phys. Rev.* **1936**, *49*, 324–328. [CrossRef]
49. Braak, D. Integrability of the Rabi Model. *Phys. Rev. Lett.* **2011**, *107*, 100401. [CrossRef] [PubMed]
50. Nataf, P.; Ciuti, C. Protected Quantum Computation with Multiple Resonators in Ultrastrong Coupling Circuit QED. *Phys. Rev. Lett.* **2011**, *107*, 190402. [CrossRef] [PubMed]
51. Romero, G.; Ballester, D.; Wang, Y.M.; Scarani, V.; Solano, E. Ultrafast Quantum Gates in Circuit QED. *Phys. Rev. Lett.* **2012**, *108*, 120501. [CrossRef] [PubMed]
52. Kyaw, T.H.; Felicetti, S.; Romero, G.; Solano, E.; Kwek, L.-C. Scalable quantum memory in the ultrastrong coupling regime. *Sci. Rep.* **2015**, *5*, 8621. [CrossRef] [PubMed]
53. Felicetti, S.; Douce, T.; Romero, G.; Milman, P.; Solano, E. Parity-dependent state engineering and tomography in the ultrastrong coupling regime. *Sci. Rep.* **2015**, *5*, 11818. [CrossRef] [PubMed]
54. Kyaw, T.H.; Herrera-Martí, D.A.; Solano, E.; Romero, G.; Kwek, L.-C. Creation of quantum error correcting codes in the ultrastrong coupling regime. *Phys. Rev. B* **2015**, *91*, 064503. [CrossRef]
55. Wang, Y.M.; Zhang, J.; Wu, C.; You, J.Q.; Romero, G. Holonomic quantum computation in the ultrastrong-coupling regime of circuit QED. *Phys. Rev. A* **2016**, *94*, 012328. [CrossRef]
56. Albarrán-Arriagada, F.; Lamata, L.; Solano, E.; Romero, G.; Retamal, J.C. Spin-1 models in the ultrastrong-coupling regime of circuit QED. *Phys. Rev. A* **2018**, *97*, 022306. [CrossRef]
57. Garziano, L.; Stassi, R.; Macrì, V.; Kockum, A.F.; Savasta, S.; Nori, F. Multiphoton quantum Rabi oscillations in ultrastrong cavity QED. *Phys. Rev. A* **2015**, *92*, 063830. [CrossRef]
58. Kockum, A.F.; Miranowicz, A.; Macrì, V.; Savasta, S.; Nori, F. Deterministic quantum nonlinear optics with single atoms and virtual photons. *Phys. Rev. A* **2017**, *95*, 063849. [CrossRef]
59. Kockum, A.F.; Macrì, V.; Garziano, L.; Savasta, S.; Nori, F. Frequency conversion in ultrastrong cavity QED. *Sci. Rep.* **2017**, *7*, 5313. [CrossRef]

60. Stassi, R.; Macrì, V.; Kockum, A.F.; di Stefano, O.; Miranowicz, A.; Savasta, S.; Nori, F. Quantum nonlinear optics without photons. *Phys. Rev. A* **2017**, *96*, 023818. [CrossRef]
61. Mlynek, J.A.; Abdumalikov, A.A., Jr.; Fink, J.M.; Steffen, L.; Baur, M.; Lang, C.; van Loo, A.F.; Wallraff, A. Demonstrating W-type entanglement of Dicke states in resonant cavity quantum electrodynamics. *Phys. Rev. A* **2012**, *86*, 053838. [CrossRef]
62. Wei, X.; Chen, M.-F. Preparation of multi-qubit *W* states in multiple resonators coupled by a superconducting qubit via adiabatic passage. *Quantum Inf. Process.* **2013**, *14*, 2419–2433. [CrossRef]
63. Liu, X.; Liao, Q.; Xu, X.; Fang, G.; Liu, S. One-step schemes for multiqubit GHZ states and W-class states in circuit QED. *Opt. Commun.* **2016**, *359*, 359–363. [CrossRef]
64. Çakmak, B.; Campbell, S.; Vacchini, B.; Müstecaplıoğlu, Ė.; Paternostro, M. Robust multipartite entanglement generation via a collision model. *Phys. Rev. A* **2019**, *99*, 012319. [CrossRef]
65. Wei, X.; Chen, M.-F. Generation of N-Qubit W State in N Separated Resonators via Resonant Interaction. *Int. J. Theor. Phys.* **2014**, *54*, 812–820. [CrossRef]
66. Egger, D.J.; Wilhelm, F.K. Multimode Circuit Quantum Electrodynamics with Hybrid Metamaterial Transmission Lines. *Phys. Rev. Lett.* **2013**, *111*, 163601. [CrossRef] [PubMed]
67. Underwood, D.L.; Shanks, W.E.; Koch, J.; Houck, A.A. Low-disorder microwave cavity lattices for quantum simulation with photons. *Phys. Rev. A* **2012**, *86*, 023837. [CrossRef]
68. Wu, Y.; Yang, X. Strong-coupling theory of periodically driven two-level systems. *Phys. Rev. Lett.* **2007**, *98*, 013601. [CrossRef]
69. Brune, M.; Raimond, J.M.; Haroche, S. Theory of the Rydberg-atom two-photon micromaser. *Phys. Rev. A* **1987**, *35*, 154. [CrossRef]
70. Brune, M.; Raimond, J.M.; Goy, P.; Davidovich, L.; Haroche, S. Realization of a two-photon maser oscillator. *Phys. Rev. Lett.* **1987**, *59*, 1899–1902. [CrossRef] [PubMed]
71. Dür, W.; Vidal, G.; Cirac, J.I. Three qubits can be entangled in two inequivalent ways. *Phys. Rev. A* **2000**, *62*, 062314. [CrossRef]
72. Macri, V.; Nori, F.; Kockum, A.F. Simple preparation of Bell and GHZ states using ultrastrong-coupling circuit QED. *Phys. Rev. A* **2018**, *98*, 062327. [CrossRef]
73. Beaudoin, F.; Gambetta, J.M.; Blais, A. Dissipation and ultrastrong coupling in circuit QED. *Phys. Rev. A* **2011**, *84*, 043832. [CrossRef]
74. Ridolfo, A.; Leib, M.; Savasta, S.; Hartmann, M.J. Photon Blockade in the Ultrastrong Coupling Regime. *Phys. Rev. Lett.* **2012** *109*, 193602. [CrossRef]
75. Settineri, A.; Macri, V.; Ridolfo, A.; di Stefano, O.; Kockum, A.F.; Nori, F.; Savasta, S. Dissipation and Thermal Noise in Hybrid Quantum Systems in the Ultrastrong Coupling Regime. *Phys. Rev. A* **2019** *98*, 053834. [CrossRef]
76. Reuther, G.M.; Zueco, D.; Deppe, F.; Hoffmann, E.; Menzel, E.P.; Weißl, T.; Mariantoni, M.; Kohler, S.; Marx, A.; Solano, E.; et al. Two-resonator circuit quantum electrodynamics: Dissipative theory. *Phys. Rev. B* **2010**, *81*, 144510. [CrossRef]
77. Forn-Díaz, P.; Romero, G.; Harmans, C.J.P.M.; Solano, E.; Mooij, J.E. Broken selection rule in the quantum Rabi model. *Sci. Rep.* **2016**, *6*, 26720. [CrossRef]
78. Boissonneault, M.; Gambetta, J.M.; Blais, A. Dispersive regime of circuit QED: Photon-dependent qubit dephasing and relaxation rates. *Phys. Rev. A* **2009**, *79*, 013819. [CrossRef]
79. Masluk, N.A. Reducing the Losses of the Fluxonium Artificial Atom. Ph.D. Thesis, Yale University, New Haven, CT, USA, 2013.
80. Yurke, B.; Denker, J.S. Quantum network theory. *Phys. Rev. A* **1984**, *29*, 1419–1437. [CrossRef]
81. Devoret, M.H. Quantum Fluctuations in electrical circuits. In *Les Houches Session LXIII*; Reynaud, S., Giacobino, E., Zinn-Justin, J., Eds.; Elsevier: Amsterdam, The Netherlands, 1997; pp. 351–386.
82. Ashhab, S.; Johansson, J.R.; Zagoskin, A.M.; Nori, F. Two-level systems driven by large-amplitude fields. *Phys. Rev. A* **2007**, *75*, 063414. [CrossRef]

symmetry

MDPI

Article

Quasiprobability Distribution Functions from Fractional Fourier Transforms

Jorge A. Anaya-Contreras [1], Arturo Zúñiga-Segundo [1] and and Héctor M. Moya-Cessa [2,*]

1 Instituto Politécnico Nacional, ESFM, Departamento de Física, Edificio 9, Unidad Profesional Adolfo López Mateos, Ciuadad de México CP 07738, Mexico; mozart13892@hotmail.com (J.A.A.-C.); azuniga@esfm.ipn.mx (A.Z.-S.)

2 Instituto Nacional de Astrofísica, Óptica y Electrónica, Calle Luis Enrique Erro 1, Santa María Tonantzintla, Puebla 72840, Mexico

* Correspondence: hmmc@inaoep.mx; Tel.: +52-222-266-3100

Received: 13 February 2019; Accepted: 5 March 2019; Published: 7 March 2019

Abstract: We show, in a formal way, how a class of complex quasiprobability distribution functions may be introduced by using the fractional Fourier transform. This leads to the Fresnel transform of a characteristic function instead of the usual Fourier transform. We end the manuscript by showing a way in which the distribution we are introducing may be reconstructed by using atom-field interactions.

Keywords: quasiprobability distribution functions; fractional Fourier transform; reconstruction of the wave function

1. Introduction

It has been already shown that quasiprobability distribution functions may be reconstructed by the measurement of atomic properties in ion-laser interactions [1] and two-level atoms interacting with quantized fields [2,3]. Such measurements of the wave function are realized usually by measuring atomic observables, namely, the atomic inversion and polarization [4–7].

Although the first quasiprobability distribution functions were introduced in the quantum realm [8–12], and are useful among other things to visualize the nonclassicality of states, for instance, the squeezing of quadratures [13,14], they may be also used to analyze classical signals [15,16].

Ideal interactions, i.e., without taking into account an environment, have shown to lead to the reconstruction of the Wigner function [3] by taking advantage of its expression in terms of the parity operator. However, the interaction of a system with its environment [17] leads to s-parametrized quasiprobability distribution functions [18–20]

$$F(\alpha,s)=\frac{2}{\pi(1-s)}\sum_{k=0}^{\infty}\left(\frac{s+1}{s-1}\right)^{k}\langle k|D^{\dagger}(\alpha)\rho D(\alpha)|k\rangle \tag{1}$$

where $D(\alpha)=\exp(\alpha a^{\dagger}-\alpha^{*}a)$, with a and $a^{\dagger}$ the annihilation and creation operators of the harmonic oscillator, respectively, is the Glauber displacement operator [21]. The state $D(\alpha)|k\rangle=|\alpha,k\rangle$ is a so-called displaced number state [22]. Note that, in order to reconstruct a given quasiprobability function it is needed to displace the system by an amplitude α and then measure the diagonal elements of the displaced density matrix.

The parameter s defines different orderings and therefore different quasiprobability distribution functions (QDF). The Glauber-Sudarshan P-function [21,23] is given for $s=1$, and is used to obtain averages of functions of *normal* ordered creation and annihilation operators; $s=-1$ gives the Husimi

Q-function, used to obtain averages of functions of *anti-normal* ordered creation and annihilation operators, while $s = 0$ is used for the *symmetric* ordering and gives the Wigner function.

Equation (1) may be rewritten as

$$F(\alpha,s) = \frac{2}{\pi(1-s)} Tr\left\{ \left(\frac{s+1}{s-1}\right)^{a^\dagger a} D^\dagger(\alpha)\rho D(\alpha) \right\}, \tag{2}$$

that, by using the commutation properties under the symbol of trace, and if the system is in a *pure* state $|\psi\rangle$, may be casted into

$$F(\alpha,s) = \frac{2}{\pi(1-s)} Tr\left\{ D(\alpha) \left(\frac{s+1}{s-1}\right)^{a^\dagger a} D^\dagger(\alpha)\rho \right\} = \frac{2}{\pi(1-s)} \langle\psi|D(\alpha) \left(\frac{s+1}{s-1}\right)^{a^\dagger a} D^\dagger(\alpha)|\psi\rangle. \tag{3}$$

Recent studies have openned the possibility of measuring, instead of observables, non-Hermitian operators [24]. It would be plausible that such measurements could be related to complex quasiprobability distributions like the McCoy-Kirkwood-Rihaczek-Dirac distribution functions [9,10,12,25].

In this contribution we would like to introduce other kind of complex quasiprobabilities that, although they could be introduced simply by taking s as a complex number, we introduce them in a formal way by considering the fractional Fourier transform (FrFT) [26–28] of a signal. Then, by writing the Dirac-delta function in terms of its FrFT, we are able to write a general expression for complex quasiprobability distributions in terms of the Fresnel transform. Indeed, the representation of these complex quasiprobability distributions in terms of a Fresnel transform implies that they are solutions of a paraxial wave equation [3]. Finally, by using an effective Hamiltonian for the atom-field interaction, we show how this quasiprobability distribution function may be reconstructed.

2. Fractional Fourier Transform

Up to a phase, the fractional Fourier Transform of a signal $\psi(x)$ can be written by the following expression [26–28]

$$\mathcal{F}_\omega\left[\psi(x)\right] = \exp\left(-i\omega\hat{a}^\dagger\hat{a}\right)\psi(x), \tag{4}$$

that may be expressed in terms of an integral transform as

$$\mathcal{F}_\omega\left[\psi(x)\right] = \int_{-\infty}^{+\infty} dx' K(x,x';\omega)\psi(x'), \tag{5}$$

where

$$K(x,x';\omega) = \frac{1}{\sqrt{2\pi i}}\sqrt{\frac{e^{i\omega}}{\sin\omega}} \exp\left[i\frac{x^2}{2}\cot\omega + i\frac{x'^2}{2}\cot\omega - ixx'\csc\omega\right]. \tag{6}$$

Then, if we consider Equation (6) as a propagator, Dirac's delta distribution function takes the form

$$\begin{aligned} \delta(x-x') &= \textstyle\int_{-\infty}^{+\infty} dx'' K(x,x'';-\omega)K(x'',x';\omega) \\ &= \tfrac{1}{2\pi\sin\omega} e^{i\frac{x'^2}{2}\cot\omega - i\frac{x^2}{2}\cot\omega} \textstyle\int_{-\infty}^{+\infty} dx'' e^{ix''(x-x')\csc\omega} \\ &= \tfrac{1}{2\pi}\exp\left[i\tfrac{x'^2}{2}\cot\omega - i\tfrac{x^2}{2}\cot\omega\right] \textstyle\int_{-\infty}^{+\infty} dx'' \exp\left[ix''(x-x')\right]. \end{aligned} \tag{7}$$

Now, if we apply the fractional Fourier transform to the Dirac delta function we obtain

$$\mathcal{F}_\omega\left[\delta(x-y)\right] = \int_{-\infty}^{+\infty} dx' K(x,x';\omega)\delta(x'-y) = K(x,y;\omega). \tag{8}$$

Then, applying the inverse fractional Fourier transform to Equation (8) we obtain an alternative representation of the Dirac delta distribution function

$$\begin{aligned}\delta(x-y) = \mathcal{F}_{-\omega}\left[\mathcal{F}_{\omega}\left[\delta(x-y)\right]\right] &= \int_{-\infty}^{+\infty} dx'' K(x,x'';-\omega)K(x'',y;\omega) \\ &= \frac{1}{2\pi}\exp\left[i\frac{y^2}{2}\cot\omega - i\frac{x^2}{2}\cot\omega\right]\int_{-\infty}^{+\infty} dx' \exp\left[ix'(x-y)\right]. \end{aligned} \tag{9}$$

From the above equation it may be seen that there is a phase multiplying the usual integral representation of the Dirac delta function, that although could be omitted by using properties of the delta function, we keep in order to obtain a quasiprobability distribution function as a fractional Fourier (Fresnel) transform of the characteristic function.

3. Probability Distribution in the Phase Space

We define $\mathcal{J}(q,p)$, a probability distribution in the phase space, as

$$\mathcal{J}(q,p) = \int_{-\infty}^{+\infty}\int_{-\infty}^{+\infty} dq'dp'\mathcal{P}(q',p')\delta(q'-q)\delta(p-p'), \tag{10}$$

and then, by using Equation (9), this distribution may be rewritten as

$$\mathcal{J}(q,p) = \frac{1}{4\pi^2}e^{i\frac{q^2}{2}\cot\alpha - i\frac{p^2}{2}\cot\beta}\int_{-\infty}^{+\infty}\int_{-\infty}^{+\infty} dudv\, e^{iup-ivq}\, Tr\left\{\hat{\rho}e^{iv\hat{q}-iu\hat{p}+i\frac{\hat{p}^2}{2}\cot\beta - i\frac{\hat{q}^2}{2}\cot\alpha}\right\}, \tag{11}$$

that because

$$e^{iv\hat{q}-iu\hat{p}+i\frac{\hat{p}^2}{2}\cot\beta-i\frac{\hat{q}^2}{2}\cot\alpha} = e^{-i\frac{u^2}{2}\tan\beta}e^{i\frac{v^2}{2}\tan\alpha}e^{iu\tan\beta\hat{q}}e^{-iv\tan\alpha\hat{p}}e^{i\frac{\hat{p}^2}{2}\cot\beta-i\frac{\hat{q}^2}{2}\cot\alpha}e^{iv\tan\alpha\hat{p}}e^{-iu\tan\beta\hat{q}}, \tag{12}$$

Equation (11) takes the form

$$\begin{aligned}\mathcal{J}(q,p) = \frac{1}{4\pi^2}e^{i\frac{q^2}{2}\cot\alpha - i\frac{p^2}{2}\cot\beta}\int_{-\infty}^{+\infty}\int_{-\infty}^{+\infty} dudv\, e^{iup-ivq}e^{-i\frac{u^2}{2}\tan\beta}e^{i\frac{v^2}{2}\tan\alpha}\times \\ \times Tr\left\{\hat{\rho}e^{iu\tan\beta\hat{q}}e^{-iv\tan\alpha\hat{p}}e^{i\frac{\hat{p}^2}{2}\cot\beta-i\frac{\hat{q}^2}{2}\cot\alpha}e^{iv\tan\alpha\hat{p}}e^{-iu\tan\beta\hat{q}}\right\}.\end{aligned} \tag{13}$$

Now, by using the equivalence

$$e^{iu\tan\beta\hat{q}}e^{-iv\tan\alpha\hat{p}} = e^{\frac{i}{2}uv\tan\alpha\tan\beta}e^{iu\tan\beta\hat{q}-iv\tan\alpha\hat{p}}, \tag{14}$$

Equation (13) may be casted into the final expression

$$\begin{aligned}\mathcal{J}(q,p) = \frac{1}{4\pi^2}e^{i\frac{q^2}{2}\cot\alpha - i\frac{p^2}{2}\cot\beta}\int_{-\infty}^{+\infty}\int_{-\infty}^{+\infty} dudv\, e^{iup-ivq}e^{-i\frac{u^2}{2}\tan\beta}e^{i\frac{v^2}{2}\tan\alpha}\times \\ \times Tr\left\{\hat{\rho}e^{iu\tan\beta\hat{q}-iv\tan\alpha\hat{p}}e^{i\frac{\hat{p}^2}{2}\cot\beta-i\frac{\hat{q}^2}{2}\cot\alpha}e^{iv\tan\alpha\hat{p}-iu\tan\beta\hat{q}}\right\}.\end{aligned} \tag{15}$$

Case $\cot\alpha = -\cot\beta = \pi$

The above quasiprobability distribution function is defined for a range of parameters α and β, however, for the sake of simplicity, we will consider the case $\cot\alpha = -\cot\beta = \pi$.

We may relate the quasiprobability distribution function $\mathcal{J}(q,p)$ to the Wigner function, by noting that, for $\cot\alpha = -\cot\beta = \pi$, Equation (15) has the form

$$\mathcal{J}(q,p) = \frac{1}{4\pi^2 i}e^{i\pi\left(\frac{p^2}{2}+\frac{q^2}{2}\right)}\int_{-\infty}^{+\infty}\int_{-\infty}^{+\infty} dudv\, e^{iup-ivq}e^{i\frac{u^2}{2\pi}+i\frac{v^2}{2\pi}}Tr\left\{\hat{\rho}e^{-i\frac{u\hat{q}}{\pi}-i\frac{v\hat{p}}{\pi}}(-1)^{\hat{n}}e^{i\frac{u\hat{q}}{\pi}+i\frac{v\hat{p}}{\pi}}\right\}. \tag{16}$$

According to trace representation of Wigner function [20]

$$\mathcal{W}\left(\frac{v}{\pi},-\frac{u}{\pi}\right) = Tr\left\{\hat{\rho}\,\frac{1}{\pi}\,e^{-i\frac{u\hat{q}}{\pi}-i\frac{v\hat{p}}{\pi}}(-1)^{\hat{n}}e^{i\frac{u\hat{q}}{\pi}+i\frac{v\hat{p}}{\pi}}\right\}, \tag{17}$$

we write the distribution $\mathcal{J}(q,p)$ as the Fresnel transform of the Wigner function

$$\mathcal{J}(q,p) = \frac{1}{4\pi i}e^{i\pi\left(\frac{p^2}{2}+\frac{q^2}{2}\right)}\int_{-\infty}^{+\infty}\int_{-\infty}^{+\infty} dudv\, e^{iup-ivq}e^{i\frac{u^2}{2\pi}+i\frac{v^2}{2\pi}}\mathcal{W}\left(\frac{v}{\pi},-\frac{u}{\pi}\right). \tag{18}$$

It is easy to show that the quasiprobability distribution (18) can be normalized

$$\begin{aligned}\int_{-\infty}^{+\infty}\int_{-\infty}^{+\infty} dqdp\,\mathcal{J}(q,p) &= \tfrac{\pi}{2}\int_{-\infty}^{+\infty}\int_{-\infty}^{+\infty} dxdy\,\left[\tfrac{1}{2\pi}\int_{-\infty}^{+\infty} dp e^{ixp}\right]\left[\tfrac{1}{2\pi}\int_{-\infty}^{+\infty} dq e^{-iyq}\right]e^{-i\frac{x^2}{2\pi}-i\frac{y^2}{2\pi}}Tr\left\{\hat{\rho}\,e^{iy\hat{q}-ix\hat{p}}\right\}\\ &= \tfrac{\pi}{2}\int_{-\infty}^{+\infty}\int_{-\infty}^{+\infty} dxdy\,\delta(x)\delta(y)e^{-i\frac{x^2}{2\pi}-i\frac{y^2}{2\pi}}Tr\left\{\hat{\rho}\,e^{iy\hat{q}-ix\hat{p}}\right\} = \tfrac{\pi}{2}Tr\left\{\hat{\rho}\right\} = \tfrac{\pi}{2}.\end{aligned} \tag{19}$$

Therefore, for normalization reasons, the quasiprobability distribution is finally given in the form

$$\mathcal{J}(q,p) = \frac{1}{4\pi^2}\int_{-\infty}^{+\infty}\int_{-\infty}^{+\infty} du\,dv\,e^{iup-ivq}e^{-i\frac{u^2}{2\pi}-i\frac{v^2}{2\pi}}Tr\left\{\hat{\rho}\,e^{iv\hat{q}-iu\hat{p}}\right\}, \tag{20}$$

that, by applying the change of variables $\beta = u/\sqrt{2}+iv/\sqrt{2}$ takes the form

$$\mathcal{J}(\alpha) = \frac{1}{2\pi^2}\int d^2\beta\,e^{\alpha\beta^*-\alpha^*\beta}e^{-\frac{i}{\pi}|\beta|^2}Tr\left\{\hat{\rho}\,\hat{D}(\beta)\right\}, \tag{21}$$

with $\alpha = q/\sqrt{2}+ip/\sqrt{2}$.

From the above expression it is direct to show that the Wigner function

$$W(\alpha) = \int d^2\beta e^{\alpha\beta^*-\alpha^*\beta}\{\rho D(\beta)\}, \tag{22}$$

and the function $\mathcal{J}(\alpha)$ may be easily related by the differential relation

$$\mathcal{J}(\alpha) = \exp\left\{\frac{i}{\pi}\frac{\partial^2}{\partial\alpha\partial\alpha^*}\right\}W(\alpha). \tag{23}$$

The above quasiprobability function may be written as a trace by noting that

$$\frac{1}{2\pi^2}\int d^2\beta\,\exp\left(-\frac{i}{\pi}|\beta|^2\right)\hat{D}(\beta) = \frac{1}{2i+\pi}\left(\frac{2i-\pi}{2i+\pi}\right)^{\hat{n}} \tag{24}$$

that leads to the trace representation of $\mathcal{J}(q,p)$

$$\mathcal{J}(q,p) = \frac{1}{2i+\pi}Tr\left\{\hat{\rho}\,\hat{D}(\alpha)\left(\frac{2i-\pi}{2i+\pi}\right)^{\hat{n}}\hat{D}^\dagger(\alpha)\right\}. \tag{25}$$

Last equation allows us to show that $\mathcal{J}(q,p)$ is correctly normalized, for this we do the double integration

$$\int_{-\infty}^{+\infty}\int_{-\infty}^{+\infty}\mathcal{J}(q,p)dqdp = Tr\left\{\hat{\rho}\frac{2}{\pi+2i}\int d^2\alpha\hat{D}(\alpha)\hat{D}^\dagger\left(\alpha e^{i\theta}\right)e^{i\theta\hat{n}}\right\} = Tr\left\{\hat{\rho}\,\hat{A}\,e^{i\theta\hat{n}}\right\}, \tag{26}$$

where we have defined

$$e^{i\theta} = \frac{2i-\pi}{2i+\pi}, \tag{27}$$

and

$$\hat{A} = \frac{2}{\pi + 2i} \int d^2\alpha \, e^{i \sin\theta |\alpha|^2} \hat{D}\left(\alpha\left(1 - e^{i\theta}\right)\right) = \frac{1}{\pi^2} \int \int d^2 z_1 d^2 z_2 \, |z_1\rangle \langle z_2| \, B(z_1, z_2, z_1^*, z_2^*) \,, \tag{28}$$

with

$$\begin{aligned} B(z_1, z_2, z_1^*, z_2^*) &= \frac{2}{\pi + 2i} \int d^2\alpha \, e^{i \sin\theta |\alpha|^2} \left\langle z_1 \left| \hat{D}\left(\alpha\left(1 - e^{i\theta}\right)\right) \right| z_2 \right\rangle \\ = \frac{2\langle z_1 | z_2 \rangle}{\pi + 2i} \int_{-\infty}^{+\infty} d\alpha_x \exp&\left(-\left(1 - e^{i\theta}\right)\alpha_x^2 + \alpha_x\left(\left(1 - e^{i\theta}\right) z_1^* - \left(1 - e^{-i\theta}\right) z_2\right)\right) \times \\ \times \int_{-\infty}^{+\infty} d\alpha_y \exp&\left(-\left(1 - e^{i\theta}\right)\alpha_x^2 + i\alpha_y\left(\left(1 - e^{i\theta}\right) z_1^* + \left(1 - e^{-i\theta}\right) z_2\right)\right) \\ = \frac{2\langle z_1 | z_2 \rangle}{\pi + 2i} \frac{\pi}{1 - e^{i\theta}} &\left[-\left(1 - e^{-i\theta}\right) z_1^* z_2\right] = \left(-\frac{|z_1|^2}{2} - \frac{|z_2 e^{-i\theta}|^2}{2} + z_1^*\left(e^{-i\theta} z_2\right)\right) \\ &\langle z_1 | e^{-i\theta\hat{n}} | z_2 \rangle \,. \end{aligned} \tag{29}$$

By replacing Equation (29) into Equation (28) we obtain

$$\hat{A} = e^{-i\theta\hat{n}} \,, \tag{30}$$

that shows that Equation (26) is correctly normalized

$$\int_{-\infty}^{+\infty} \int_{-\infty}^{+\infty} \mathcal{J}(q, p) dq dp = Tr\left\{\hat{\rho} \, e^{-i\theta\hat{n}} \, e^{i\theta\hat{n}}\right\} = Tr\left\{\hat{\rho}\right\} = 1 \,. \tag{31}$$

4. Kirkwood Distribution and $\mathcal{J}(q, p)$ Distribution

The Kirkwood distribution is defined as [12,25,29,30]

$$\mathcal{K}(q, p) = \frac{1}{4\pi^2} \int_{-\infty}^{+\infty} \int_{-\infty}^{+\infty} du \, dv \, e^{iup - ivq} e^{i\frac{uv}{2}} \, Tr\left\{\hat{\rho} e^{iv\hat{q} - iu\hat{p}}\right\} \,, \tag{32}$$

or an alternative way to write it as an expectation value [31] is

$$\mathcal{K}(q, p) = \frac{1}{\sqrt{2}\pi} e^{\frac{q^2}{2} + \frac{p^2}{2} + iqp} \left\langle -i\sqrt{2}p \left| e^{\frac{\hat{a}^2}{2}} \hat{\rho} e^{-\frac{\hat{a}^{\dagger 2}}{2}} \right| \sqrt{2}q \right\rangle . \tag{33}$$

The Kirkwood function belongs to a class of QDFs that although is complex, still has the same amount of information as other real QDFs, namely Wigner, Glauber Sudarshan or Husimi distribution functions.

Being the QDF $\mathcal{J}(q, p)$ and Kirkwood distributions complex functions we show now some differences between them.

4.1. Number State

The Kirkwood $\mathcal{K}(q, p)$ and $\mathcal{J}(q, p)$ distributions for number state $|n\rangle$, are represented by the following equations

$$\mathcal{K}_n(q, p) = \frac{i^n}{2^n n! \pi\sqrt{2}} e^{-\frac{q^2}{2} - \frac{p^2}{2} + iqp} H_n(q) H_n(p) \tag{34}$$

and

$$\mathcal{J}_n(q, p) = \frac{1}{2i + \pi}\left(\frac{2i - \pi}{2i + \pi}\right)^n \exp\left(-\frac{\pi(q^2 + p^2)}{2i + \pi}\right) L_n\left(\frac{2\pi^2(q^2 + p^2)}{4 + \pi^2}\right) \,, \tag{35}$$

where, $H_n(x)$ and $L_n(x)$ are Hermite and Laguerre polynomials, respectively.

4.2. Superposition of Two Coherent States

Now, we consider a superposition of two coherent states as:

$$|\psi_\pm\rangle = \frac{1}{\sqrt{2 \pm 2\mathrm{Re}\,\langle\alpha_1|\alpha_2\rangle}} \left(|\alpha_1\rangle \pm |\alpha_2\rangle\right), \tag{36}$$

where $\alpha_k = q_k/\sqrt{2} + ip_k/\sqrt{2}$, such that the Kirkwood $\mathcal{K}(q,p)$ and the $\mathcal{J}(q,p)$ distributions for the superposition of two coherent states, $|\psi_\pm\rangle$, is given by

$$\begin{aligned}
\mathcal{J}_\pm(q,p) = {} & \frac{1}{2i+\pi}\frac{1}{2\pm 2\mathrm{Re}\,\langle\alpha_1|\alpha_2\rangle}\left(\exp\left(-\frac{\pi}{2i+\pi}\left((q-q_1)^2+(p-p_1)^2\right)\right)\right) \\
& \frac{1}{2i+\pi}\frac{1}{2\pm 2\mathrm{Re}\,\langle\alpha_1|\alpha_2\rangle}\left(\exp\left(-\frac{\pi}{2i+\pi}\left((q-q_2)^2+(p-p_2)^2\right)\right)\right) \\
\pm & \frac{1}{2i+\pi}\frac{1}{2\pm 2\mathrm{Re}\,\langle\alpha_1|\alpha_2\rangle}\left(\exp\left(\frac{i}{2}\left(q\left(p_1-p_2\right)-p\left(q_1-q_2\right)\right)\right)\left\langle\alpha_2-\alpha\middle|e^{i\theta}(\alpha_1-\alpha)\right\rangle\right) \\
\pm & \frac{1}{2i+\pi}\frac{1}{2\pm 2\mathrm{Re}\,\langle\alpha_1|\alpha_2\rangle}\left(\exp\left(-\frac{i}{2}\left(q\left(p_1-p_2\right)-p\left(q_1-q_2\right)\right)\right)\left\langle\alpha_1-\alpha\middle|e^{i\theta}(\alpha_2-\alpha)\right\rangle\right)
\end{aligned} \tag{37}$$

and

$$\begin{aligned}
\mathrm{K}_\pm(q,p) = {} & \frac{1}{\sqrt{2}\pi}\frac{\exp\left(-\frac{q^2}{2}-\frac{p^2}{2}+iqp\right)}{2\pm2\mathrm{Re}\langle\alpha_1|\alpha_2\rangle}\left(\exp\left(-\tfrac{1}{2}(q_1^2+p_1^2)+(qq_1-pp_1)+\tfrac{i}{2}q_1(p_1+2p)+\tfrac{i}{2}p_1(q_1-2q)\right)\right) \\
+ & \frac{1}{\sqrt{2}\pi}\frac{\exp\left(-\frac{q^2}{2}-\frac{p^2}{2}+iqp\right)}{2\pm2\mathrm{Re}\langle\alpha_1|\alpha_2\rangle}\left(\exp\left(-\tfrac{1}{2}(q_2^2+p_2^2)+(qq_2-pp_2)+\tfrac{i}{2}q_2(p_2+2p)+\tfrac{i}{2}p_2(q_2-2q)\right)\right) \\
\pm & \frac{1}{\sqrt{2}\pi}\frac{\exp\left(-\frac{q^2}{2}-\frac{p^2}{2}+iqp\right)}{2\pm2\mathrm{Re}\langle\alpha_1|\alpha_2\rangle}\left(\exp\left(-\tfrac{1}{2}(q_1^2+p_2^2)+(qq_1-pp_2)+\tfrac{i}{2}q_2(p_2+2p)+\tfrac{i}{2}p_1(q_1-2q)\right)\right) \\
\pm & \frac{1}{\sqrt{2}\pi}\frac{\exp\left(-\frac{q^2}{2}-\frac{p^2}{2}+iqp\right)}{2\pm2\mathrm{Re}\langle\alpha_1|\alpha_2\rangle}\left(\exp\left(-\tfrac{1}{2}(q_2^2+p_1^2)+(qq_2-pp_1)+\tfrac{i}{2}q_1(p_1+2p)+\tfrac{i}{2}p_2(q_2-2q)\right)\right),
\end{aligned}$$

respectively.

We plot both distribution in Figures 1 and 2. In both figures a more uniform behaviour may be seen in the QDF $\mathcal{J}_\pm(q,p)$ than in the Kirkwood function. In fact, the real and imaginary parts of the distribution we have introduced here, look like Wigner function for number states (Figure 1) and Scrhödinger cat states (Figure 2).

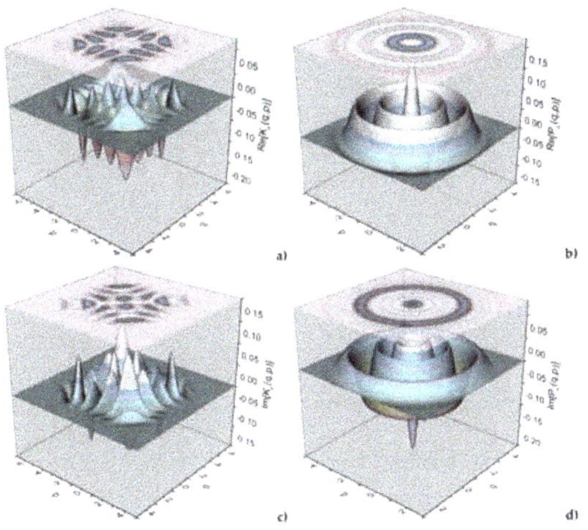

Figure 1. In figures (**a**,**c**) we can see the phase space distribution of the real and imaginary parts of the Kirkwood function for a number state $|n = 3\rangle$. In figures (**b**,**d**) we see the distribution $\mathcal{J}(q, p)$, for the same number state, again, the real and imaginary parts, respectively.

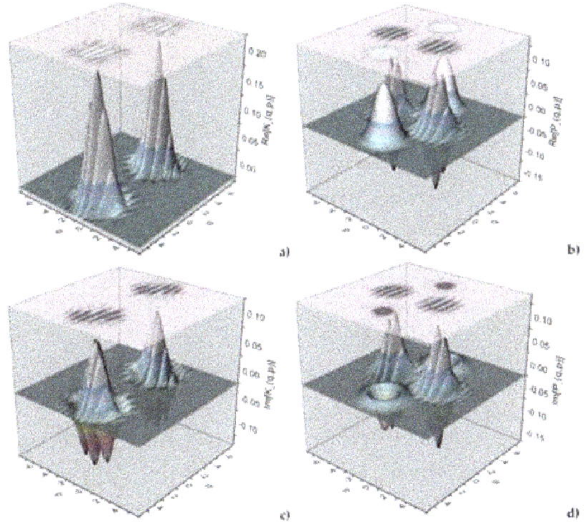

Figure 2. In figures (**a**,**c**) we can see the phase space distribution of the real and imaginary parts of the Kirkwood function for two superposition of coherent states $|\psi_+\rangle$ wiht $q_1 = -q_2 = 4$ and $p_1 = p_2 = 0$. In figures (**b**,**d**) we see the distribution $\mathcal{J}(q, p)$, again, the real and imaginary parts, respectively.

5. Reconstruction of Distribution $\mathcal{J}(\alpha)$

It is not difficult to show that the real part of QDF $\mathcal{J}(\alpha)$ may be measured. This can be achieved by measuring the atomic polarization in the dispersive interaction between an atom and a quantized field [3], whose Hamiltonian reads

$$H = -\chi a^{\dagger} a \sigma_z, \tag{38}$$

with $\sigma_z = |e\rangle\langle e| - |g\rangle\langle g|$, the Pauli matrix corresponding to the atomic inversion operator, where $|g\rangle$ and $|e\rangle$ represent the ground and excited states of the two-level atom. The parameter χ is the dispersive coupling constant. The above Hamiltonian yields the evolution operator

$$U(t) = \exp\{-i\chi t a^{\dagger} a \sigma_z\}, \tag{39}$$

from which we can obtain the evolved wavefunction $|\psi(t)\rangle = U(t)|\psi(0)\rangle$, that allows the calculation of averages of different observables.

The average of observable $\sigma_x = |e\rangle\langle g| + |g\rangle\langle e|$ then can be obtained for an arbitrary initial field, which we conveniently write as $|\psi_F(0)\rangle = \hat{D}^{\dagger}(\alpha)|\phi(0)\rangle$ and the atom is initially in a superposition of atomic states, $|\psi_A(0)\rangle = \frac{1}{\sqrt{2}}(|g\rangle + |e\rangle)$. Then we write

$$\langle\sigma_x(t)\rangle = \frac{1}{2}\left(\langle\phi(0)|\hat{D}(\alpha)\exp\{2i\chi t a^{\dagger}a\}\hat{D}^{\dagger}(\alpha)|\phi(0)\rangle + c.c.\right). \tag{40}$$

Of course, if in this equation we set $t = \pi/(2\chi)$, we would recover the Wigner distribution function [3,18,32,33], as

$$\left\langle \sigma_x\left(\frac{\pi}{2\chi}\right)\right\rangle = \langle\phi(0)|\hat{D}(\alpha)\cos\{\pi a^{\dagger}a\}\hat{D}^{\dagger}(\alpha)|\phi(0)\rangle, \tag{41}$$

is proportional to the s-parametrized quasiprobability distribution function of Equation (2) for $s = 0$ [1–3,32,33].

It is also easy to show that the imaginary part of the QDF may be associated to the observable $\sigma_y = i(|e\rangle\langle g| - |g\rangle\langle e|)$

$$\langle\sigma_y(t)\rangle = \frac{i}{2}\left(\langle\phi(0)|\hat{D}(\alpha)\exp\{2i\chi t a^{\dagger}a\}\hat{D}^{\dagger}(\alpha)|\phi(0)\rangle - c.c.\right). \tag{42}$$

If we set the interaction time $t = \frac{\arctan\frac{4\pi}{\pi^2-4}}{2\chi}$, we obtain that

$$Re\{\mathcal{J}(\alpha)\} \propto \langle\sigma_x\rangle, \qquad Im\{\mathcal{J}(\alpha)\} \propto \langle\sigma_y\rangle. \tag{43}$$

Therefore, by measuring the polarizations σ_x and σ_y we are able to measure the QDF $\mathcal{J}(\alpha)$.

6. Conclusions

We have introduced a set of parametrized (in terms of α and β) quasiprobability distribution functions, Equation (15), by using the fractional Fourier transform. This has lead us to generalize QDF to Fresnel transforms of the characteristic function instead of their usual Fourier transforms. We have also shown how such QDF may be reconstructed in the dispersive atom-field interaction. We have also given a (differential) relation that allows the calculation of the newly introduced QDF from the Wigner function.

Finally, we would like to stress that the distribution function we are introducing may be of importance in problems in which non-Hermitian operators are measured.

Author Contributions: J.A.A.-C. conceived the idea and developed it under A.Z.-S. and H.M.M.-C. supervision. The manuscript was written by all authors, who have read and approved the final manuscript.

Funding: This research received no external funding.

Acknowledgments: We thank CONACYT for support.

Conflicts of Interest: The authors declare no conflict of interest.

References

1. Leibfried, D.; Meekhof, D.M.; King, B.E.; Monroe, C.; Itano, W.M.; Wineland, D.J. Experimental determination of the motional quantum state of a trapped atom. *Phys. Rev. Lett.* **1996**, *77*, 4281. [CrossRef] [PubMed]
2. Bertet, P.; Auffeves, A.; Maioli, P.; Osnaghi, S.; Meunier, T.; Brune, M.; Raimond, J.M.; Haroche, S. Direct measurement of the Wigner function of a one-photon fock state in a cavity. *Phys. Rev. Lett.* **2002**, *89*, 200402. [CrossRef] [PubMed]
3. Lutterbach, L.G.; Davidovich, L. Method for direct measurement of the Wigner function in cavity QED and ion traps. *Phys. Rev. Lett.* **1997**, *78*, 2547. [CrossRef]
4. Leonhardt, U. *Measuring the Quantum State of Light*; Cambridge University Press: Cambridge, UK, 1997.
5. Lvovsky, A.I.; Raymer, M.G. Continuous-variable optical quantum-state tomography. *Rev. Mod. Phys.* **2009**, *81*, 299–332. [CrossRef]
6. Wallentowitz, S.; Vogel, W. Reconstruction of the quantum-mechanical state of a trapped ion. *Phys. Rev. Lett.* **1995**, *75*, 2932–2935. [CrossRef] [PubMed]
7. Wallentowitz, S.; Vogel, W. Unbalanced homodyning for quantum state measurements. *Phys. Rev. A* **1996**, *53*, 4528–4533. [CrossRef] [PubMed]
8. Wigner, E.P. On the quantum correction for thermodynamic equilibrium. *Phys. Rev.* **1932**, *40*, 749. [CrossRef]
9. McCoy, N.H. On the function in quantum mechanics which corresponds to a given function in classical mechanics. *Proc. Natl. Acad. Sci. USA* **1932**, *18*, 674. [CrossRef] [PubMed]
10. Dirac, P.A.M. On the analogy between classical and quantum mechanics. *Rev. Mod. Phys.* **1945**, *17*, 195. [CrossRef]
11. Husimi, K. Some formal properties of the density matrix *Proc. Phys. Math. Soc. Jpn.* **1940**, *22*, 264–314.
12. Kirkwood, J.G. Quantum statistics of almost classical assemblies. *Phys. Rev.* **1933**, *44*, 31–37. [CrossRef]
13. Kiesel, T.; Vogel, W.; Bellini, M.; Zavatta, A. Nonclassicality quasiprobability of single-photon-added thermal states. *Phys. Rev. A* **2011**, *83*, 032116. [CrossRef]
14. Moya-Cessa, H.; Vidiella-Barranco, A. Interaction of squeezed states of light with two-level atoms. *J. Mod. Opt.* **1992**, *39*, 2481–2499. [CrossRef]
15. Alonso, M.A. Wigner functions in optics: Describing beams as ray bundles and pulses as particle ensembles. *Adv. Opt. Photonics* **2001**, *3*, 272–365. [CrossRef]
16. Bastiaans, M.J.; Wolf, K.B. Phase reconstruction from intensity measurements in linear systems. *J. Opt. Soc. Am. A* **2003**, *20*, 1046–1049. [CrossRef]
17. Yazdanpanah, N.; Tavassoly, M.K.; Jurez-Amaro, R.; Moya-Cessa, H.M. Reconstruction of quasiprobability distribution functions of the cavity field considering field and atomic decays. *Opt. Commun.* **2017**, *400*, 69–73. [CrossRef]
18. Royer, A. Wigner function as the expectation value of a parity operator. *Phys. Rev. A* **1977**, *15*, 449. [CrossRef]
19. Wünsche, A. Displaced Fock states and their connection to quasi-probabilities. *Quantum Opt.* **1991**, *3*, 359–383. [CrossRef]
20. Moya-Cessa, H.; Knight, P.L. Series representation of quantum-field quasiprobabilities. *Phys. Rev. A* **1993**, *48*, 2479. [CrossRef] [PubMed]
21. Glauber, R.J. Coherent and incoherent states of the radiation field. *Phys. Rev.* **1963**, *131*, 2766. [CrossRef]
22. De Oliveira, F.A.M.; Kim, M.S.; Knight, P.L.; Buzek, V. Properties of displaced number states. *Phys. Rev. A* **1990**, *41*, 2645. [CrossRef] [PubMed]
23. Sudarshan, E.C.G. Equivalence of semiclassical and quantum mechanical descriptions of statistical light beams. *Phys. Rev. Lett.* **1963**, *10*, 277. [CrossRef]
24. Pati, A.K.; Singh, U.; Sinha, U. Measuring non-Hermitian operators via weak values. *Phys. Rev. A* **2015**, *92*, 052120. [CrossRef]
25. Rihaczek, A.N. Signal energy distribution in time and frequency. *IEEE Trans. Inf. Theory* **1968**, *14*, 369–374. [CrossRef]
26. Namias, V. The fractional order Fourier transform and its application to quantum mechanics. *J. Inst. Math. Appl.* **1980**, *25*, 241–265. [CrossRef]
27. Agarwal, G.S.; Simon, R. A simple realization of fractional Fourier transform and relation to harmonic oscillator Green's function. *Opt. Commun.* **1994**, *110*, 23. [CrossRef]

28. Fan, H.-Y.; Chen, J.-H. On the core of the fractional Fourier transform and its role in composing complex fractional Fourier transformations and Fresnel transformations. *Front. Phys.* **2015**, *10*, 100301. [CrossRef]
29. Praxmeyer, L.; Wódkiewicz, K. Quantum interference in the Kirkwood-Rihaczek representation. *Opt. Commun.* **2003**, *223*, 349–365. [CrossRef]
30. Praxmeyer, L.; Wódkiewicz, K. Hydrogen atom in phase space: The Kirkwood-Rihaczek representation. *Phys. Rev. A* **2003**, *67*, 054502. [CrossRef]
31. Moya-Cessa, H. Relation between the Glauber-Sudarshan and Kirkwood-Rihaczec distribution functions. *J. Mod. Opt.* **2013**, *60*, 726–730. [CrossRef]
32. Moya-Cessa, H.; Roversi, J.A.; Dutra, S.M.; Vidiella-Barranco, A. Recovering coherence from decoherence: A method of quantum state reconstrucion. *Phys. Rev. A* **1999**, *60*, 4029–4033. [CrossRef]
33. Moya-Cessa, H.; Dutra, S.M.; Roversi, J.A.; Vidiella-Barranco, A. Quantum state reconstruction in the presence of dissipation. *J. Mod. Opt.* **1999**, *46*, 555–558. [CrossRef]

MDPI
St. Alban-Anlage 66
4052 Basel
Switzerland
Tel. +41 61 683 77 34
Fax +41 61 302 89 18
www.mdpi.com

Symmetry Editorial Office
E-mail: symmetry@mdpi.com
www.mdpi.com/journal/symmetry

www.ingramcontent.com/pod-product-compliance
Lightning Source LLC
Chambersburg PA
CBHW051916210326
41597CB00033B/6166

* 9 7 8 3 0 3 9 2 1 8 5 8 5 *